THE FOREST FRONTIER

THE FOREST FRONTIER

Settlement and Change in Brazilian Roraima

Edited by Peter A. Furley

London and New York

First published 1994
by Routledge
11 New Fetter Lane, London EC4P 4EE
Simultaneously published in the USA and Canada
by Routledge
29 West 35th Street, New York, NY 10001

Typeset in Garamond by J&L Composition Ltd, Filey, North Yorkshire
Printed and bound in Great Britain by Biddles Ltd, Guildford and King's Lynn

British Library Cataloguing in Publication Data
A catalogue record for this book is available from the British Library

Library of Congress Cataloging in Publication Data
has been applied for

ISBN 0–415–04392–1

CONTENTS

CONTENTS

ILLUSTRATIONS

PLATES

FIGURES

TABLES

CONTRIBUTORS

Christopher Barrow. Lectures in the Centre for Development Studies, University College, Swansea.

Tom Dargie. Independent Environmental Consultant, formerly Lecturer in the Department of Geography, University of Sheffield.

Michael Eden. Lectures in Geography at Royal Holloway & Bedford New College, University of London.

Peter Furley. Reader in Tropical Soils and Biogeography, University of Edinburgh, formerly Professor of Ecology (Soils) in the University of Brasilia.

John Hemming. Director and Secretary of the Royal Geographical Society, London; leader of the RGS/INPA/SEMA Maracá Rain Forest Project and author of several books on Brazil.

Philippe Léna. Works for ORSTOM (the French Overseas Agency) and is based at the Museu Goeldi in Belém, Pará.

Gordon MacMillan. Spent 16 months in Roraima researching informal sector mining for his PhD at the University of Edinburgh.

Duncan McGregor. Senior Lecturer in Geography at Royal Holloway & Bedford New College, University of London.

Luc Mougeot. Works for the International Development Research Centre (IDRC), Ottawa and is Senior Lecturer on leave from the Federal University of Pará, Belém.

Andrew Paterson. Consultant specialising in horticulture and based at Bradford upon Avon in England.

PREFACE

This book evolved from research carried out in conjunction with the RGS/ INPA/SEMA Maracá Rain Forest Project (Hemming 1988, 1989). Although it was not possible to assemble a team large enough to examine the issues of environment and development comprehensively, a number of central themes were covered over a two to three year period from 1986 with later fieldwork from 1991 to 1992. Remarkably little has been published, even in Portuguese, on the nature and resources of the region in relation to developments such as the change from territory to state or the impact of the gold-rush. What is available frequently has a relatively restricted distribution. Consequently, the book attempts to bring together original research along with an overview of resources and an assessment of recent developments.

Roraima has many environmental and human features which make up a unique combination within Brazil. In addition to being the most northerly and, until recently, one of the most isolated parts of Brazil, it shares international boundaries with two neighbouring countries, Venezuela and Guyana. It contains the only substantial area of savanna in Brazil lying north of the Amazon. Furthermore, although it shares, with much of the rest of the Brazilian Amazon, a characteristic flat, wet forested cover, it nevertheless also possesses spectacular and distinctive mountains and residual hills. These lie to the north and west bordering Venezuela and Guyana and are characterised by the dramatic peak of Mt Roraima shared between the three countries. The state has one of the highest concentrations of Indians in Brazil, including one of the largest groups, the Yanomami. Their lands formed the focus of the recent gold strikes which exceeded many of the better-known gold-rushes of the nineteenth and twentieth centuries. For a period of around three years, gold dominated the economic and social life of the state and has left a perceptible imprint on its rural and urban population.

The aims of the book may be simplified under three headings: firstly, to outline the intrinsic environmental and socio-economic resources and to consider their potential; secondly, to assess current land developments as a means of assisting in the planning process and, thirdly, to highlight selected aspects of this process which formed part of the Maracá Rain Forest Project and are therefore mostly concerned with the northern parts of the state.

In the first chapter, Peter Furley and Luc Mougeot explain the context of the book by means of an overview of the physical character and socio-economic issues facing the new federal state. This is reinforced by John Hemming's chapter dealing with the emergence of Roraima in its current political form. This inevitably focusses upon the early explorations and overriding influence of ranching. The following chapter by Tom Dargie and Peter Furley considers the problems of survey and monitoring of resources in poorly known and inaccessible humid tropical areas and pays particular attention to the region between the capital, Boa Vista westwards and beyond Maracá Island. The chapter by Michael Eden and Duncan McGregor concentrates on the process of deforestation and pasture substitution and considers the problems of land degradation.

An important aspect of development has been the process of land occupation – colonisation schemes and spontaneous settlement. This is dealt with in the study by Luc Mougeot and Philippe Léna examining a series of such developments to the west and north-west of Boa Vista. By contrast, Christopher Barrow and Andrew Paterson assess the use of some of the most agriculturally promising lands in the region – the *várzeas* or riverine plains. In the final chapter Gordon MacMillan and Peter Furley explore some of the implications of current developments, particularly the impact of the gold discoveries, and examine whether Roraima can avoid the same pressures and problems that concern environmentalists and social planners in the southern Amazonian states.

ACRONYMS

ASTER	Associação de Assistência Técnica e Extensão Rural (Roraima)/Association for Technical Assistance and Rural Extension
AVHRR	Advanced Very High Resolution Radar
BASA	Banco da Amazônia SA/Bank of Amazônia
CCPY	Comissão pela Criação do Parque Indigena Ianomami
CEDI	Centro Ecumênico de Documentação e Informação/Ecumenical Centre for Documentation and Information
CNPH	Centro Nacional de Pesquisa de Hortaliças/ National Centre for Horticultural Research
CPAC	Centro de Pesquisa Agropecuária do Cerrado/ Centre for Agropastoral Research in the Cerrado (part of EMBRAPA)
CPRM	Companhia de Pesquisa do Recursos Minerais (technical wing within DNPM)
DNPM	Departamento Nacional de Pesquisas Minerais/ National Department of Mineral Research (Geological Exploration)
EMATER	Empresa de Assistência Técnica e Extensão Rural/ Agency for technical assistance and agricultural extension
EMBRAPA	Empresa Brasileira de Pesquisa Agropecúaria/ Brazilian agency for agro-pastoral research
FUNAI	Fundação Nacional do Indio/National Indian Foundation
IBAMA	Instituto Brasileiro do Meio Ambiente/Brazilian Environmental Institute
IBDF	Instituto Brasileiro de Desenvolvimento Florestal/ Brazilian Forestry Institute (now part of IBAMA)

IBGE	Instituto Brasileiro de Geografia e Estatística/ Brazilian Institute of Geography and Statistics
INCRA	Instituto Nacional de Colonização e Reforma Agraria/National Institute of Colonisation and Agrarian Reform
INPE	Instituto de Pesquisas Espaciais/The Brazilian Space Agency
PAD	Projeto de Assentamento Dirigido/Directed Settlement Project
PIC	Projeto Integrado de Colonisação/Integrated Colonisation Project
POLOAMAZÔNIA	Programa de Polos Agropecuários e Agrominerais da Amazônia/Programme of Agro-Pastoral and Agro-Mineral Poles in Amazonia
PROVAM	Programa de Estudos e Pesquisas dos Vales Amazônico
SEMA	Secretaria Especial do Meio Ambiente/Special Secretariat for the Environment (now part of IBAMA)
SLAR	Side Looking Airborne Radar
SUCAM	Superintendência da Campanha de Saude Pública de Amazônia/Agency for the Public Health Campaign for Amazonia
SUDAM	Superintendência de Desenvolvimento da Amazônia/Agency for the development of Amazonia

1

PERSPECTIVES

Peter Furley and Luc Mougeot

INTRODUCTION

The development of Roraima, the most northerly of the Brazilian Amazon states, which stretches from just south of the Equator to over 5° North, is of particular interest at the present time. On the one hand, it mirrors many of the current and familiar problems of the humid tropics but at a lesser stage of advancement; yet on the other, the region has a character and individuality quite distinct from the rest of Brazil and justifies a separate focus of attention. With an area of little over 230,000 sq. km (88,800 sq. miles) and a population of under 250,000 (IBGE 1992), it is evident that socio-economic change in Roraima, though rapid through the 1980s, has been at a less frenetic pace than that typical of the southern Brazilian Amazon. An arc of development, driven by demographic and economic forces from the south, extends from Acre in the west through Rondônia, Mato Grosso, Pará, Tocantins to Maranhão in the east (Figure 1.1).

Amongst the most urgent of the development issues are the competitive pressures on land and the destruction of natural vegetation; the conflicts between indigenous groups and colonisers; the environmental and social power of the great mining companies and smaller scale but potent effects of *garimpeiros* (miners or prospectors); the search for sources of energy and an eruption of proposals for hydroelectric power schemes; the road building, land speculation and land conflicts. These characteristic southern Amazonian features are compounded by a remarkable Brazilian pioneering spirit which began with the *bandeirantes* (explorers) and are unlikely to fade before all the national territory is occupied. The context of development and processes associated with the fluxes of people and capital are well documented elsewhere (for example Barbira-Scazzochio 1980; Moran 1981,83; Hemming 1985; Fearnside 1986a; Hecht and Cockburn 1989; Hall 1989; Goodman and Hall 1990; Eden 1991; Cleary 1987, 1991).

The problems associated with development south of the Basin have spread infectiously throughout the whole of the Amazon (*Amazônia Legal*, Figure 1.1), wherever communications have aided the movement of people (Bourne 1978). Despite the pressures, the Brazilian territory lying to the north of the

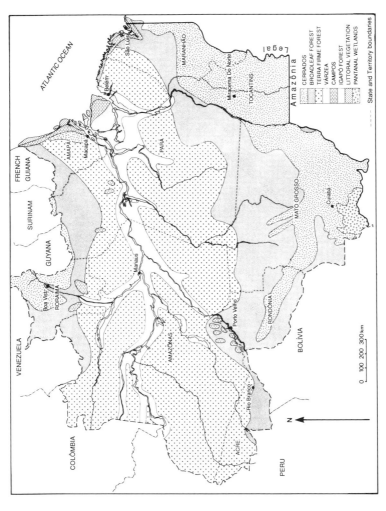

Figure 1.1 The states to the north of the Amazon river, from Amapá (to the east) through Pará, Roraima and Amazonas form the focus of the vast, 6,500-km-long, internal and external security zone known as the Calha Norte. With the southern states from Acre (to the west) through Rondônia, Mato Grosso, Tocantins and Maranhão, they make up the administrative unit known as Amazônia Legal. Note that much of the basin is covered by savanna (*cerrado*), grassland (*campo*), or wetlands (*várzeas* along periodically flooded river plains and *pantanal* to the south-west).

river remains so far relatively unscathed, despite the military 'inroads' resulting from the *Calha Norte* project (Pacheco 1990; Albert 1992; Allen 1992) and mining in parts of Amapá, northern Amazonas and Roraima. Little is known of Roraima even within Brazil, and a remarkably small amount of information was available as background to the transition from Federal territory to state in 1990. Even many of the texts dealing with the Amazon scarcely contain a reference to Roraima. This does not imply that Roraima has escaped from the development pressures but, in the latter part of the 1980s and early 1990s, it has remained very much a frontier area (Aubertin 1988; Schmink and Wood 1985, 1992; Hemming 1985).

ENVIRONMENTAL AND BIOLOGICAL RESOURCES

Of the Brazilian Amazonian states, Roraima, and to a lesser extent Amapá, possesses the most substantial tracts of savanna (*cerrado*) lying to the north of the river. It also has a larger proportion of hills and mountains to the north and west than most of the Basin (Figure 1.2). Although dominated by forest, the tree cover is not all evergreen but becomes partially deciduous and broken towards the north as the savanna boundaries are approached and in places where the soils are richer. The pattern of vegetation is mainly a result of the more seasonally dry conditions found to the north-east. However, with a substantial river system largely coincident with the catchment basin of the Rio Branco, the state is well supplied with water for most of its area. There are rich botanic and faunal resources and a diversity of soils which, though generally poor in terms of agricultural potential, have at least a significant usable proportion. Along with considerable mineral reserves especially of gold and cassiterite, these characteristics provide the region with a range of physical and biological resources. In sum, the environmental resources of Roraima will be shown to be both similar to and yet uniquely different from the Brazilian Amazon as a whole (Prance and Lovejoy 1985; Goodland and Irwin 1975; Dickinson 1987; Prance 1982; Jordan 1987; HMSO 1991).

Whereas the northern and western parts of Roraima are rolling and at times mountainous, culminating in the summit of Mount Roraima (2,772 metres), the greater part of the state is low-lying with little relief. The accidented topography and deep valleys with fast-flowing rivers full of rapids is characteristic of the ancient north and west of the country. The level wet plains are typical of the much more recent alluvial deposits. The state capital, Boa Vista, lies over 2,000 km up river from the Amazon estuary and yet has an elevation of less than 100 metres above sea level. The panorama southwards from the spectacular residual hills (*tepuis*) of northern Roraima, usually made up of sandstones overlying the Guiana Complex (Figure 1.6), looks out over a broad, level plain with an upper basin around Boa Vista distinguished from the southern Rio Branco plains, separated by

3

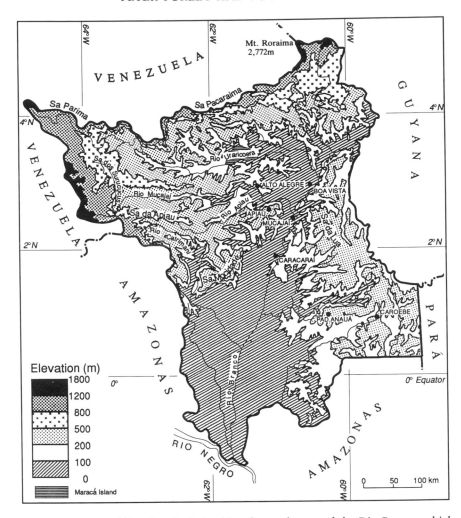

Figure 1.2 Most of Roraima is drained by the catchment of the Rio Branco which connects to the Rio Negro and thence to the Amazon below Manaus.

a slight topographic constriction around Caracaraí, the second town of the state.

Climatic resources

Much of the natural resource potential of Roraima ultimately depends on climate. Unfortunately, data on climate resources are extremely sparse with the exception of the old established meteorological station at Boa Vista. Recent climatic assessments have been based on the stations at Boa Vista and Barcellos (Amazonas state) representing the north-east and southern parts of Roraima respectively. From the little information available, it is evident that, apart from the hilly areas, rainfall increases southwards and westwards from the dry north-east corner (Figure 1.3). Boa Vista has a total annual rainfall of less than 2,000 mm and 7 dry months with less than 100 mm per month. The rainfall variability also increases towards the north-east and the nature and duration of the wet season can be critical for cultivation and ranching. Most of the forested areas would appear to have totals in the order of 1,750 to over 2,000 mm and the rainfall increases westwards, as do the number of rainy months in the year. Effective humidity reflects altitude as well as seasonality, as illustrated in Figure 1.4.

Over the lowland areas, the data suggest a mean annual temperature of around 26 to 27°C with a small mean monthly range of 2–3°C. Temperatures are rarely limiting, although they are likely to be lower over the highest mountains and obviously relate to evapotranspiration rates – which show deficits to the north-east corner of the state. The estimated index of thermal efficiency reflects the smaller values associated with uplands and the increase towards the Equator (Figure 1.5). Disturbance to the vegetation, particularly deforestation, is likely to accentuate seasonal water deficiencies and may have severe impacts on the environment (Salati et al. 1986).

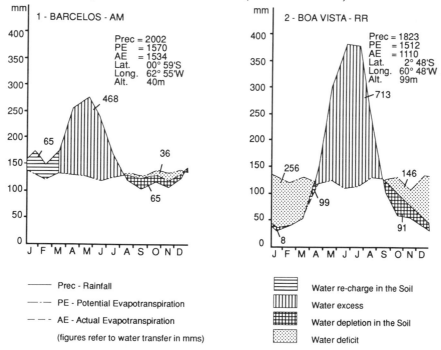

Figure 1.3 Climatic data for Barcelos and Boa Vista.
Source: After IBGE 1981, *Atlas de Roraima*.

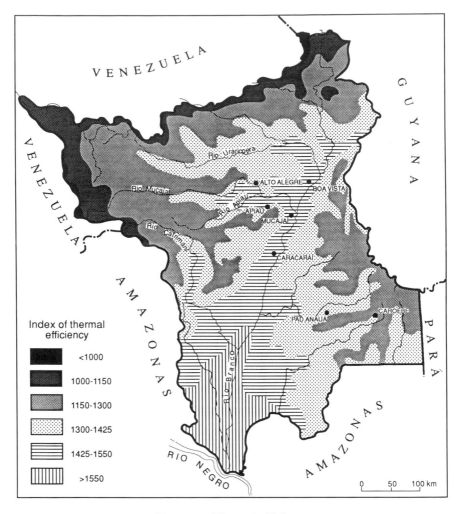

Figure 1.4 Thermal efficiency.
Source: After IBGE 1981, *Atlas de Roraima*, and Brasil 1978, Projeto Radam.

Geology and geomorphology

The greater part of the state is underlain by an early Cambrian contorted crystalline platform known as the Guiana Complex. This is made up of the Basal Shield with metamorphic gneisses and mica schists and equally old intrusive granodiorite rocks (Figure 1.6). The Shield areas are well referred to as a Complex, not only on account of their varied mineral composition, but because they have suffered numerous (at least 5 major) episodes of tectonic upheaval with intermediate periods of volcanic activity, often

6

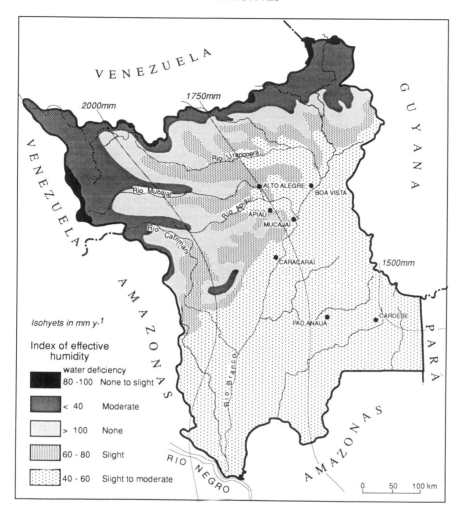

Figure 1.5 Effective humidity.
Source: After IBGE 1981, *Atlas de Roraima.*

leaving distinctive 'stocks' in the landscape. Towards the north and west of the state, the platform is covered by mid and late Precambrian rocks which make up the present-day mountainous landscape. In places, they form the residual hills which are composed of sandstones of the Roraima Group. Although such features are better developed in adjacent areas of southern Venezuela they occur locally in Roraima as at Mount Roraima, Tepequém north of Maracá Island, or in the Serras further to the west. The mineral composition of the Shield rocks varies from andesites and rhyolites in the Surumu Formation to the north, to the more granitic rocks of the Surucucu

7

The significance of the geology from the point of view of land development in Roraima lies in its potential for mining exploitation, in its influence over topography and drainage and in its parent role in soil formation. Mineral resources were first identified in the 1920s by Oliveira (1929, 1937), but the importance of diamonds and gold amongst other mineral resources had to wait over 20 years before they were revealed in the more detailed surveys of Barbosa and Ramos (1959). The only comprehensive, though reconnaissance investigation, has been the survey undertaken by the geological division of Projeto Radambrasil (Brasil 1975a, 1975b, 1978; Ramgrab *et al.* 1972).

A number of important mineral resources have been exploited or have the potential to influence development. Diamonds were discovered in the Maú river on the Guyanan border at the beginning of the century. Later on, deposits were mined in the Suapi and Quinô rivers draining into the Rio Cotingo in the far north-east, as well as the better known but isolated Tepequém site (Oliveira *et al.* 1969) (see Figure 1.14). Several other diamond-rich areas have been discovered and the quality seems to improve westwards; therefore informal mining (*garipagem*) conflicts with the area of predominantly Indian population (Brasil 1975a). The same is true for tin, which has been discovered in the Yanomami areas to the north-west and in the Waimiri–Atroari Reserve (Figure 2.1) in the south close to the BR 174. Tin has also stimulated a major development, informally by the gold-miners, particularly in the north-western Surucucus region (Figure 1.2) and originally as a discarded by-product, and formally in the major Pitinga mine on the border between Amazonas and Roraima. Gold has been a major influence, particularly during the recent gold-rush period from 1987 to 1990 (see Chapter 7). At times gold is found in the same locations as diamonds (such as Fazenda Mina Seca and upper reaches of the Maú and Cotingo rivers). Both alluvial deposits and primary rock formations have been mined, although the placer (alluvial) areas are the ones most commonly exploited by *garimpeiros*. Other minerals which have either been developed to a slight degree or are likely to assume greater importance in the future include further deposits of tin (cassiterite) to the east of the state, agate, ilmenite and columbite, iron, aluminium and nickel conglomerates, manganese, sulphates, barytes, and rare, often dispersed minerals such as zircon, titanium, molybdenum and radioactive minerals.

The landscapes of Roraima have developed over many cycles of rejuvenation and planation. This has left a series of undulating or dissected surfaces sometimes sharply separated by escarpments. The older, higher remnants at around 500 to 600 metres above sea level (asl), are today mostly fragmented in the north or exist as isolated blocks further south, such as the Serras do Demini and da Lua (Figure 1.2). They comprise the interfluvial *planaltos* or elevated tablelands between the Orinoco and Amazon drainage catchments and the residual surfaces depicted in Figure 1.7. These planation levels have

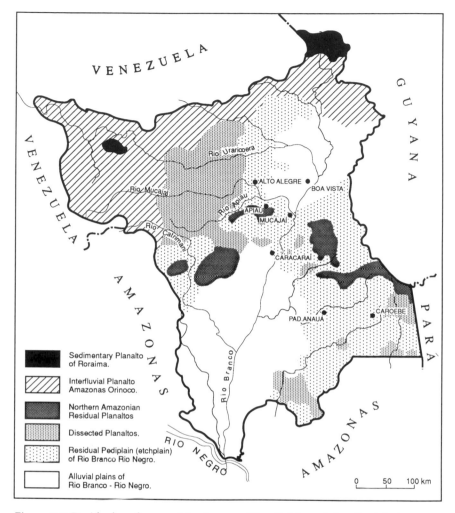

Figure 1.7 Residual surfaces and landscapes. The division of the alluvial plains into the dry north-eastern section centred on Boa Vista, and the extensive swampy southern section, is caused by constriction due to older and higher land between Caracaraí and Mucajaí.

Source: IBGE 1981, modified from *Atlas de Roraima*.

been matched across the Guiana Shield and correspond with the Oligocene mid-Tertiary surface in the middle Amazon (Bigarella and Ferreira 1985). A lower and more extensive surface, at about 250–400 metres asl, remains to the west of Boa Vista as an isolated fragment and more continuously to the south and east (known as the Northern Amazonian Dissected Planalto). This conforms with a general late Tertiary (upper Miocene/lower Pliocene) surface. The most widespread surface lies at some 80–160 metres asl and has

11

been covered by deposits from the late Tertiary (the residual pediplains or etchplains of the Rio Branco and Negro, Figure 1.7). This also accords with the widespread Amazonian Plio-Pleistocene surface lying between 50 and 200 metres asl (Bigarella and Ferreira 1985). Finally, the most recent incisions into this surface follow the contemporary drainage and make up the present day alluvial flood plains. The drainage pattern is curious in its sharp westward change of direction in the Rio Uraricoera (or Urariquera) north of Boa Vista. It is believed that this results from the capture in late Tertiary by the Rio Branco of the originally eastward-flowing Uraricoera and Mucajaí, which formed a separate catchment and would have flowed into Guyana and thence to the Atlantic (McConnell 1968; Eden and McGregor 1989). This is supported by the division of the 80–160 metre surface into north and south sections by a residual band of older and higher land extending across the Caracaraí region (Figures 1.2 and 1.7). The details of the geomorphology are largely speculative in view of the reconnaisance scales of existing surveys.

Soils and land evaluation

The only comprehensive survey of soil resources was carried out as part of Projeto Radam Brasil (Brasil 1975a, 1975b, 1978; Almeida 1984; Furley 1986), but the soils were later re-evaluated for land assessment purposes (Brasil 1980). This generated maps at 1:1 m from field survey based on SLAR (Side Looking Airborne Radar) using 1:250,000 mosaics as base maps. The density of sampling was inevitably light given the pioneering nature and reconnaissance scale of the surveys, but there is agreement with later work at a broad level.

The main groups of soil are the Latosols (Oxisols) and Red-Yellow Podzolic soils (Ultisols), characteristic not just of Roraima but of the Brazilian Amazon (Furley 1990) and the Amazon Basin as a whole (Cochrane et al. 1985) (see Figure 1.8). The Latosols are deep, intensely weathered, highly leached and usually very porous soils. They vary greatly in terms of physical properties such as texture and colour, and in their chemical attributes, which are controlled by parent materials and degree of development. Most soils can be characterised as dystrophic – that is, acidic with low nutrient reserves and exchange capacity. They occur in the middle reaches of northern rivers mostly associated with level and gently undulating relief, and under a variety of vegetation types from dense forest to open savanna. The only soils with good potential for agriculture are the Dark Red Latosols, particularly the sub-group *Terra Roxa Estruturada* (Alfisol, US Soil Taxonomy) to the west of Boa Vista, which have finer textures but are nevertheless well drained and deep. However, this group of richer soils only covers a limited area (less than 5 per cent) and the majority of Roraima's surface consists of the acid, nutrient-deficient soils typical of most of the

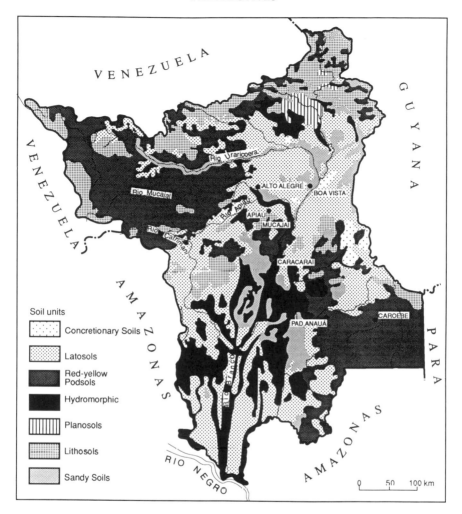

Figure 1.8 Distribution of soil types. The soil units follow the terminology of the Brazilian Soil Survey, SNLCS (Serviço Nacional de Levantamento e Conservação do Solo).

Amazon Basin. The Podzolic soils are found along the basins of the larger rivers in areas of undulating and strongly rolling relief up to 1,000 metres asl. They are clay-rich below sandy topsoils, acidic, highly susceptible to weathering and erosion and are not usually considered appropriate for current agricultural practices.

Other important groups include hydromorphic, planosol, lithosol, concretionary, lateritic and quartz sand/alluvial soils. As might be predicted, the hydromorphic group are found along the *várzeas* (riverine plains) and *igapós* (regularly inundated forest and swamplands). They typically contain

13

organic surface horizons and are frequently gleyed. They are often associated with lateritic horizons where hardening of illuviated iron occurs during periods of drying and oxidation. Planosols can be considered as a variant group of the hydromorphic soils, found on the planation surfaces described earlier. Such soils may accumulate a clay-rich and impermeable sub-surface and are predominantly very wet. Lithosols are shallow soils with only surface soil development, occurring over rocky outcrops and always closely related to the parent material. They are found on steep slopes and over the higher Serras such as Parima. Concretionary and lateritic soils are also thin and unsuited to any form of agricultural development. They form a very varied group with agglomerated nodules or masses of iron-rich material and are especially concentrated to the east of Boa Vista in the upper Tacutu Basin (Figure 1.14). The lateritic soils tend to be associated with ground water but may have accumulated colluvially in past drier periods. Finally, the quartz-rich and alluvial soils occupy a large area of the state associated with current and ancient stream lines. Over 95 per cent are made up of sand, mostly quartzitic, but containing heavy minerals such as zircon, tourmaline and rutile. Their coarse textures, extreme acidity, high porosity and leaching combine to produce soils of very low fertility.

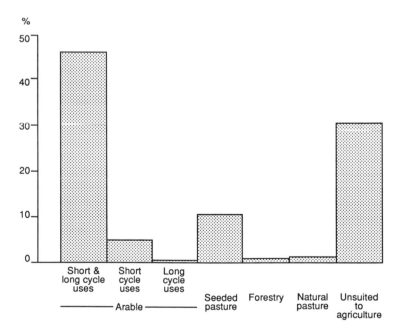

Figure 1.9 Land evaluation classes. A very small proportion of land is suitable for forestry and for ranching on natural pastures; 47 per cent of the area is considered appropriate for short- to long-cycle crops, though this has yet to be demonstrated in field experiments; a very large area is designated as unsuited to any form of agriculture (under present conditions and knowledge).

On the basis of the soil and land aptitude surveys of Projeto Radam, the Ministry of Agriculture later re-assessed the agricultural potential for Roraima (Brasil 1980). It will be evident from the environmental considerations dealt with so far, that constraints to agricultural development are considerable. Nevertheless the land evaluation estimated that over 57 per cent of the state was capable of supporting short- and long-cycle arable farming, assuming the use of appropriate fertilisers (Figure 1.9). The land aptitude appraisal reported that, in 1980, around 10 per cent of this area was at that time apportioned to parks and reserves. Seeded pastures made up a further 11 per cent and only a small area (c.1 per cent) could be used sustainably under natural pasture. A tenth of this group existed as parks and reserves. Only a small area was suited to commercial forestry on account of low growth rates (around 14,000 ha with high productivity of > 150 m³3/ha). Not surprisingly, given the limitations outlined, a large proportion of the state (43 per cent) was considered unsuitable for unrestricted agricultural use.

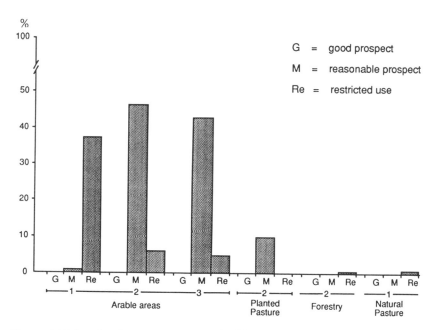

Figure 1.10 Levels of management required for potential land use. Agricultural suitability is shown at different levels of management (excluding parks and reserves). Cultivated lands at subsistence levels of management (1) are very restricted; with intermediate levels of technology (2) and advanced management (3), reasonable potentials are envisaged; sown pastures at the intermediate level, have some prospect, but even with the higher levels of management, the use of land for natural pastures or forestry is exceedingly limited given present knowledge.
Source: Based on Brasil, SUPLAN 1980.

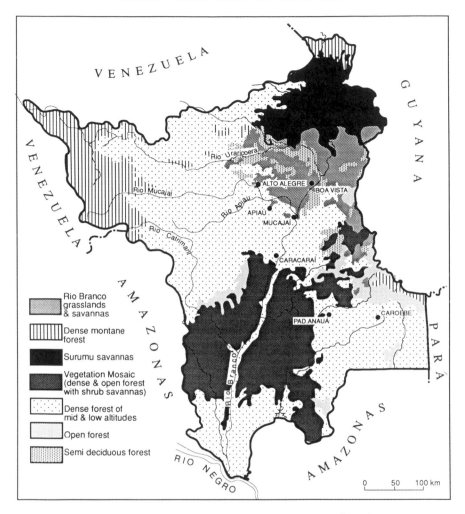

Figure 1.12 Vegetation classes. Map excludes swampy and semi-permanent inundated tracts within forest zones or seasonally flooded areas on the *cerrado*. *Source*: IBGE 1981, modified from *Atlas de Roraima*.

pioneer formations and ecological tension zones by the Radam surveys). Differences in soil reflecting topographic, textural, nutrient and water controls, affect both the morphology and floristic character of vegetation over short distances.

The *cerrados* occupy the area of marked seasonality with a dry season lasting up to six months. They are continuous with the southern Venezuelan llanos and Guyanese Rupununi savannas (Eden 1974). In addition to the seasonally dry water courses and depressed river levels, there are numerous discrete ovoid depressions which are also water-filled in wet periods. The intervening mounds are likely to be residual, left by differential water erosion and many of the dry mounds are subsequently occupied by termites giving rise to a characteristic 'pock-marked' landscape (*campos de murundus*) (Diniz *et al.* 1986; Oliveira and Furley 1990), although other areas may be dominated by termite activity. As in many other Brazilian savanna regions, it is often difficult to distinguish natural from anthropogenic influences in separating the open arboreal and shrubby *cerrados* from the nearly treeless grasslands.

There is a gradient of *cerrado* sub-groups from the frequently sharp forest–savanna boundary (see examples in Furley *et al.* 1992), through a fairly dense arboreal savanna which becomes progressively more open and shrubby, to a nearly homogeneous herbaceous sward (Goodland and Ferri 1979; Eiten 1972; Furley and Ratter 1988). Sometimes this reflects a hydrological catena, with better-drained forested upslope conditions followed by a zone of woody *cerrado* lying above seasonally flooded areas with hyperseasonal grassy savanna (Sarmiento 1984).

Disturbance to the natural vegetation has not been as severe as in many southern Amazon states, but there are, nevertheless, significant signs of change in the land cover. This topic is treated in more detail in Chapter 3, which examines the rates of change from satellite data, and Chapter 4, which presents the results of field surveys examining the process of forest transformation to pasture. Recent information for the whole state was collected by IBDF, currently IBAMA (Brasil 1983) which monitors forest destruction. Although Figure 1.13 is several years out of date it does show the nuclei of disturbance which can be seen to relate to (a) urban expansion and influence of Boa Vista, (b) the road network notably along the BR 174 with offshoots towards the Guyanese border and along the unfinished northern perimeter road and (c) associated with land colonisation schemes such as Alto Alegre (Chapters 3 and 5) and around Pad. Anauá and along the BR 210 towards Caroebe (Figure 1.14).

SETTLEMENT AND RESOURCE DEVELOPMENT

Any account of human settlement and resource development needs to have an appreciation of historical evolution and an understanding of the stimuli and constraints to progress. During most of its history (see Chapter 2), Roraima has remained remote from Brazilian markets and population centres. At the same time, its people have had to learn how to exploit the region's comparatively greater access to cross-border trade and migration routes. The state has more extensive international (1,922 km) than national (1,535 km) borders. Within Brazil, Roraima is often perceived as distant and remote, yet it is also – to a surprising, if embryonic extent – a cosmopolitan frontierland. Both in the capital and on interior estates, it is not unusual to meet hotel and restaurant workers, farm managers, labourers or miners who

Figure 1.13 Forest clearance and initial colonisation in the early 1980s.
Source: After Brasil, IBDF 1983.

speak Spanish, English, Surinamese or Indian languages in addition to, even sometimes with little, Portuguese. Imported goods are common in urban stores and in small riverside or roadside shops, such as whisky or cement from Britain and Canada (via Guyana), butter from Surinam (from Europe), or bullets and beer from Central America (through Venezuela).

Perceptions of Roraima

If there are few people outside Brazil who know or have even heard of Roraima, there are many misconceptions about the state from Brazilians who live further south. They find it difficult to imagine the distance and isolation. For example, Boa Vista lies further west than Vilhena in Rondônia, further north than the whole of Amapá, and is 600 km north of Manaus by direct air connection. Until recently, it was quicker to get to Georgetown, Guyana, or Paramaribo in Surinam than to Manaus. Flight times from Boa Vista to Belém are longer than to Porto Velho (970 km) or Caracas (785 km), partly because the journey is indirect. Boa Vista is, in fact, the state capital most distant from Brasília; more than Rio Branco in Acre or even Fernando de Noronha 650 km off Ceará's seaboard.

Roraima is substantial in size by European standards, being little smaller than Great Britain and two and a half times that of the colonial parent, Portugal. It is also significant by US standards with an area between that of Utah and Oregon, and only Alaska and ten other US states are larger. Although approximately equal to Rondônia, São Paulo state or Piaui, Roraima still represents only 3 per cent of Brazilian territory and a little under 8 per cent of Amazônia Legal. Its area per capita places it amongst the least densely settled places in the habitable world. Despite this, Roraima is comparatively well developed in relation to parts of north-west Amazonas, but it has only one major road, the BR 174 running from Venezuela in the north to join the Rio Negro at Manaus. This has a few offshoots to the Guyanese border (BR 401), fragments of the ill-fated northern perimeter road (BR 210), and local networks of dirt trails to colonisation schemes and pioneer ranches (see Figure 1.14). It is indeed a frontierland, even if the traditional notions of frontiers are now less relevant for the Brazilian Amazon as a whole; it is literally a frontier for agricultural penetration and mining exploration, biologically a frontier between forest and savanna, a political frontier of growing importance and, perhaps most telling, a far-off frontier in the perception of most Brazilians.

In this initial overview of the socio-economic aspects of development, emphasis is given to the official statistics. Despite this apparent substance to the evidence, it should be remembered that there is considerable doubt as to the accuracy, completeness and quality of the data, and that recent evidence has not been systematically collected. The figures should be treated with caution and taken as the best available trends.

Historical and political evolution

Along with much of the Amazon basin, there has been a quickening pace of development in Roraima, particularly over the last one or two decades. It is this rapidity of change which fuels anxieties over the long-term stability of development and which makes Roraima of special interest. The details of exploration, discovery and settlement are examined later but it is worth remembering the physical and mental barriers to penetration and communication faced by the early Portuguese explorers and their Indian guides

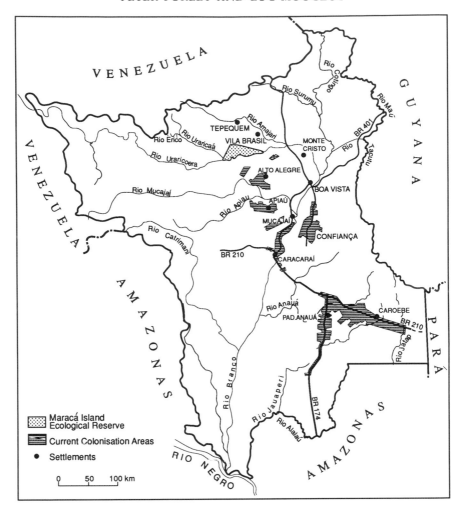

Figure 1.14 Road system and the main colonisation settlements.

as they followed the waterways into what is now Roraima, or the sheer magnitude of distance faced by the present-day Brazilian land settlers.

It is not difficult to understand why the Portuguese conquest of Roraima was intermittent and why the integration of the state into mainstream Brazilian development has been slow and incomplete. Even today, the demarcation of parts of the border remains to be settled. The early Portuguese land claims, the missionary endeavours of the Jesuits, the clashes with Dutch, English and Spanish incursions are considered in the next chapter, but there have been continuous and in many ways unsuccessful attempts to utilise natural resources from the beginning – along with

Figure 1.15 Municipalities in Roraima in the 1980s and 1990s.

political aspirations for expansion. The political existence of Roraima is fairly recent and will almost certainly undergo considerable changes with further population growth, infrastructural development and expansion of national and international trade. From a position similar to that of a 'parish' in 1858 to that of a small town in 1890, Boa Vista became a municipality in 1892. With this municipality and parts of others in Amazonas state, the territory of Roraima was created half a century later. Since 1943, the territory has been progressively sub-divided into further municipalities. There were eight of them in 1988, when the territory was finally elevated into a full state (Figure 1.15) and there are plans to create more.

23

The role of ranching and transport in development

Since the earliest colonial settlement, the Upper Rio Branco economy, like that of the natural grasslands and seasonal wetlands of Marajo at the mouth of the Amazon, has revolved around cattle ranching (Rivière 1972). This has proved to be the most reliable way of raising and protecting capital in these distant lands (Hecht 1985). Even today, cattle are used as a means of avoiding depreciation of savings, guaranteeing and/or refunding bank loans, cancelling personal debts, defraying children's education and treating health, purchasing property as well as paying for durable goods and labour or, in times of scarcity, for basic food staples. Recently, however, there have been signs that traditional ranching – extensive grazing on natural grasslands – is approaching the limits of natural environmental carrying capacity and is encroaching into forest areas (Chapters 3 and 4; Eden et al. 1991).

Table 1.2 Number of cattle and areas under pasture, 1970–87: Brazil, The North Region and Roraima

	Cattle (heads)	Pastures (hectares)	Ratio
Brazil			
1970	78,453,143	145,138,529	1.96
1975	102,531,778	165,652,250	1.62
1980	118,971,418	174,499,641	1.47
1970–80 % change	51.65	13.21	−25.00
1985	128,422,666		
1987	135,726,280		
North			
1970	1,709,890	4,428,116	2.59
1975	2,113,448	5,281,440	2.50
1980	3,687,747	7,722,487	2.09
1979–80 % change	115.67	74.40	−19.31
1985	5,273,372		
1987	6,899,166		
Roraima			
1970	237,550	1,147,034	4.83
1975	249,978	1,353,168	5.41
1980	326,097	1,601,784	4.91
1970–80 % change	37.28	39.65	1.66
1985	305,155		
1987	362,386		

Sources: IBGE (1974, 1979, 1984, 1988) *Anuários Estatísticos do Brasil* 1973, 1978, 1983, 1987/8, 1989, pp. 165; 388; 390; 431–2; 436; 351, 336 (Rio de Janeiro: IBGE); IBGE (1975) *VIII Recenseamento Geral – 1970. Censo Agropecuário Brasil*, Série Nacional, vol. III, pp. 440–1 (Rio de Janeiro: IBGE); IBGE (1979) *Censos Econômicos de 1975. Censo Agropecuário Brasil*, Série Nacional, vol. 1, pp. 246–7 (Rio de Janeiro: IBGE); IBGE (1984) *IX Recenseamento Geral – 1980. Censo Agropecuário Brasil*, vol. 2, tomo 3, no. 1, pp. 264–5 (Rio de Janeiro: IBGE); IBGE (1988) *Anuário Estatístico do Brasil 1987/8*, p. 311 (Rio de Janeiro: IBGE). (Note that these are mostly estimates.)

One indication of this is the recent evolution of Roraima's cattle herds. In the 1970s and early 1980s they grew at a considerably slower rate than in the North Region as a whole (Table 1.2). In fact the participation of Roraima in the regional total fell from 1970 to 1987. Reports suggest a marked decline in cattle numbers since the gold-rush of the late 1980s (see Figure 2.2; Chapter 7). The problem is not only numbers but also quality and access to markets. In 1985, the market value of Roraima's beef was less than 60 per cent of the regional average.

The comparatively slow expansion of the Roraima cattle population is at least partly a reflection of the decreasing carrying capacity of the upper Rio Branco natural grasslands. This has been accentuated by increasing competition from more intensive land uses in the upper Rio Branco, particularly in the vicinity of Boa Vista. Whilst cattle density per hectare of pasture grew by 25 per cent in Brazil and 20 per cent in the North Region, the density in Roraima stayed around the same over the decade 1970–80 (Table 1.2). In particular, around Boa Vista, extensive irrigated grain crops on seasonally flooded *várzeas*, are now competing with traditional ranching, both for land and critical water supplies. This trend is likely to increase in the future as a larger consumer market grows in the capital and road links to the Manaus region improve. Furthermore, other more productive ranching systems are developing elsewhere in the Amazon and the products might well penetrate and increasingly supply Roraima's traditionally captive markets. Beef is already being imported from other parts of Brazil.

Land transportation network

In this delicately poised situation, evidence is emerging of changes already taking place in an attempt to improve the state's productive performance. A new phase has been initiated in the settlement and resource exploitation of the region, with a spatial reach and pace unprecedented in Roraima's history. This new momentum is being fuelled by two processes: an expanding land transportation network and a growing influx of people from other Brazilian states.

In terms of land transportation, during the period 1970–87 when national and regional systems expanded their roadlengths by 42 and 160 per cent respectively, the embryonic Roraima network grew by 355 per cent (Table 1.3). Road building activity in Roraima was particularly intense during the period 1970–75 when, every week for 5 years, 3.6 km of new road were added to the system – mostly linking Boa Vista to Manaus.

Changes in the control of Roraima's road system are also taking place and could markedly affect further land development. While 70 per cent of the total road system was under Federal control in 1970, the situation was reversed by 1987, on the brink of independent statehood, with 62 per cent being under Territorial jurisdiction. The conversion of the territory into a

25

Table 1.3 Roadlength open to traffic and distribution according to federal, state and municipal jurisdiction, 1970–88

	All	Federal (%)			State (%)			Municipal (%)		
	Total*	Total	Paved	Unpaved	Total	Paved	Unpaved	Total	Paved	Unpaved
Brazil										
1970	1,039,779	4.96	2.32	2.63	12.05	2.25	9.80	82.99	0.16	82.83
1975	1,428,602	4.83	2.81	2.01	7.93	1.49	6.44	87.24	0.22	87.02
1980	1,371,598	4.59	2.89	1.42	9.35	2.47	6.32	86.06	0.43	85.63
1985	1,417,248	4.55	3.28	1.02	10.58	3.65	6.05	84.87	0.44	84.38
1987	1,473,003	4.42	3.30	0.93	10.35	3.95	6.28	84.73	0.61	84.12
1988	1,502,594	4.41	3.29	0.89	12.50	4.93	6.92	83.09	0.67	82.41
North										
1970	21,260	20.90	1.23	19.68	38.90	6.90	32.00	40.19	0.81	39.38
1975	28,361	34.45	5.45	28.99	18.91	5.09	13.82	46.89	1.39	45.50
1980	41,688	25.72	4.09	20.07	24.29	4.61	18.44	49.99	2.23	47.75
1985	44,116	24.76	5.73	17.17	25.68	5.00	19.55	49.55	1.92	47.46
1987	55,343	20.42	4.73	14.13	19.82	4.13	14.44	59.77	0.30	59.33
1988†	80,890	15.12	4.17	9.89	24.20	3.61	19.23	60.67	1.08	59.50
Roraima										
1970	720	69.72	–	69.72	–	–	–	30.28	0.42	29.86
1975	1,664	65.38	–	65.38	–	–	–	34.62	0.18	34.44
1980	2,499	50.90	1.04	45.26	49.10	–	43.78	–	–	–
1985	3,202	38.54	1.25	37.29	61.46	–	61.46	–	–	–
1987	3,279	37.85	1.74	33.36	62.15	–	58.65	–	–	–
1988	3,373	36.82	1.69	32.43	63.18	–	59.77	–	–	–

Sources: IBGE (1972, 1978, 1984, 1988) *Annuários Estatísticos do Brasil* 1971, 1977, 1983, 1987/8, 1989, pp. 398; 573–4; 636–7; 669–79 and 633, respectively (Rio de Janeiro: IBGE).
Notes: * Roadlength in kilometres; road sections under repair excluded from percentages.
† Includes Tocantins.

state and the creation of new municipalities will probably extend the road system as it has done elsewhere in Brazil and help to decentralise decision-making on road development. The Roraima road network has clearly only just begun to develop and it has been argued that there is considerable need, as well as room, for expansion. Theoretically, at least, there should be a greater opportunity to benefit from past experience in the assessment of road construction plans and projects than has been the case in most other parts of Amazonia. This is particularly important, considering that much development is taking place in the sensitive environments of the southern and western forest areas.

Population growth and settlement

The second major process which sustains Roraima's development, is its population growth. In the late 1970s and 1980s, Roraima more or less repeated the population boom experienced around a decade earlier by Rondônia but to a lesser scale (Tables 1.4 and 1.5). Comparatively, however, Roraima's record makes a deeper impression, on account of the state's extreme isolation from major migrant-supplying regions and agriculture-consumer markets, as well as less spectacular official propaganda for colonisation schemes.

Despite the isolation, population growth is largely attributable to sustained inward migration from the late 1970s. The main influx of migrants is quite recent when compared with national and regional figures (Table 1.6). The share of migrants having at least 10 years residence was much smaller in 1980 than in the North Region or Brazil as a whole. In Roraima, 75 per cent arrived during the 1970s. By 1980, one-third of all residents had been born in a different state, giving an idea of Roraima's improving net attractiveness for both Roraima natives and those born elsewhere in Brazil. The number of Roraima-born residents in other states has shown a progressive decline from 1950 to 1980. Migrant flows are aimed primarily at the few urban centres, which in the 1970s boomed at rates well above national and regional

Table 1.4 Population – National, North Region, Rondônia and Roraima, 1950–89

	Brazil	North Region	Rondônia	Roraima
1950	51,944,397	1,844,655	36,935	18,116
1960	70,070,457	2,561,782	69,792	28,304
1970	93,139,037	3,603,860	111,064	40,885
1980	119,002,706	5,880,268	491,069	79,159
1989	50,051,784	9,925,301*	1,021,229	130,070

Source: Reproduced from IBGE (1988) *Anuário Estatístico do Brasil 1987/8, 1989 Suplemento* p. 11, p. 58 (Rio de Janeiro: IBGE).
Note: * Includes Tocantins.

Table 1.5 Changes in population density (inhabitants per km^2) in National, North Region and constituent states/territories, 1940–88

	1940	1950	1960	1970	1980	1988
Brazil	4.88	6.14	8.29	11.01	14.07	17.63
North Region	0.41	0.52	0.72	1.01	1.65	2.58
Roraima	–	0.08	0.13	0.18	0.34	0.57
Rondônia	–	0.15	0.29	0.46	2.02	4.28
Acre	0.52	0.75	1.04	1.41	1.97	2.68
Amazonas	0.28	0.33	0.45	0.61	0.92	1.37
Pará	0.77	0.92	1.25	1.77	2.77	4.01
Amapá	–	0.27	0.49	0.82	1.26	1.67
Tocantins	–	–	–	–	–	3.48

Source: Reproduced from IBGE (1988) *Anuário Estatístico do Brasil 1987/8*, p. 94 (Rio de Janeiro: IBGE).

averages (Table 1.7). This is certainly having considerable socio-cultural and politico-economic effects on local and traditional ways of life. Such contrasts are especially noticeable in Boa Vista, which dominates the population and where most outsiders tend to reside (see Figure 2.3).

One extremely interesting trend which became visible in the 1980 Census relates to the new position that Roraima was assuming in the 'search space' of people moving to Amazonia. By 1980 it had become quite clear that Roraima received growing numbers of people who proceed directly from their native states located outside the North Region. Maranhão provides a major share and, with Amazonas, Pará and Rio Grande do Sul, accounted for 62 per cent of all direct migrants in the 1970s. This is also true of states held up until recently as promising frontiers of settlement. A comparison of migrant origins reveals that Rondônia, Mato Grosso, Goias (including Tocantins) and the Federal District are now serving as intermediate stations for a growing number of migrants. After some time spent in these states south of the Amazon they have moved in search of new or better opportunities. This may reflect government incentives for grain production, the prospect of obtaining land or, later, the lure of *garimpagem*.

Land development and land tenure

A combination of feeder-road and track expansion and inter-state migration has promoted a still largely spontaneous and uncontrolled form of settlement. Nevertheless, colonisation projects have been either Federal or state planned, and most of the population increase up to the mid-1980s was essentially 'planned'. This has resulted in the incorporation of large and mostly forested tracts of land into the rural tenure system. What has happened in Roraima in the recent past needs to be appreciated within the broader national and regional context (Table 1.8). During 1970–85, there was not much change on the Brazilian scene: the total area of rural estates

Table 1.6 Population born outside *município* of current residence and percentage distribution according to period of residence 1970 and 1980: Brazil, North Region and Roraima

		Period of residence in years								
	Total	Less than 1	1	2	3	4	5	6–9	10 or more	Undeclared
Brazil										
1970	30,270,451	11.66	5.99	7.25	5.84	4.68	4.80	18.51*	41.24†	0.02
1980	46,342,494	10.65	5.96	6.50	5.75	4.86	4.64	13.06	48.21	0.39
North										
1970	547,933	11.95	5.79	7.04	6.11	4.81	5.53	17.85	40.86	0.05
1980	1,778,453	13.69	8.41	8.54	8.05	7.01	5.99	15.09	32.71	0.52
Roraima										
1970	3,967	17.04	10.01	9.07	5.77	4.46	7.26	18.53	27.68	0.18
1980	25,212	15.92	10.89	11.46	10.29	5.62	7.01	13.77	24.71	0.32

Sources: OBGE (1975) *VIII Recenseamento Geral – 1970. Censo Demográfico Brasil*, Série Nacional, vol. 1, pp. 139–40 (Rio de Janeiro: IBGE); IBGE (1983) *IX Recenseamento Geral – 1980. Censo Demográfico Brasil*, vol. 1, tomo 4, no. 1, pp. 102–6 (Rio de Janeiro: IBGE).

Notes: * 1970 published census figures correspond to a six-to-ten-year period.
† 1970 published census figures correspond to a period of eleven years or more.

Table 1.7 Mean annual intercensal population growth, 1940–80

	1940–50	1950–60	1960–70	1970–80
Brazil				
Total	2.39	2.99	2.89	2.48
Urban	3.91	5.15	5.22	4.44
Rural	1.60	1.55	0.57	0.62
North				
Total	2.29	3.34	3.47	5.02
Urban	3.71	5.04	5.44	6.44
Rural	1.84	2.37	2.11	3.70
Roraima				
Total	5.49	4.65	3.75	6.83
Urban	–	8.84	3.71	10.80
Rural	–	2.17	3.78	2.66

Source: Reproduced from IBGE *Anuário Estatístico do Brasil 1987/8*, p. 94 (Rio de Janeiro: IBGE).

Table 1.8 Total area in rural estates and percentage distribution according to tenancy status of producer, in Brazil, the North Region and Roraima, 1970–85

	All	Tenancy status			
	Total	Proprietor	Leaseholder	Sharecropper	Occupant
Brazil					
1970	293,012,168	86.03	4.81	1.79	7.37
1975	323,896,082	89.16	3.00	0.96	6.89
1980	369,587,872	86.32	3.95	1.50	7.19
1985	376,286,577	87.89	3.45	1.69	5.35
North					
1970	22,482,707	55.46	15.45	3.07	26.02
1975	32,615,964	68.05	4.82	0.50	26.63
1980	42,546,027	61.53	7.39	0.77	28.64
1985*	–	–	–	–	–
Roraima					
1970	1,587,406	89.65	2.41	2.45	5.49
1975	1,836,406	18.16	0.00	0.98	80.85
1980	2,478,767	31.93	0.12	0.26	67.14
1985*	–	–	–	–	–

Sources: IBGE (1974, 1980, 1984, 1988) *Anuários Estatísticos do Brasil 1973, 1979, 1983, 1987/88*, pp. 168; 320; 376; and 310–11, respectively (Rio de Janeiro: IBGE).
Note: * Preliminary results unavailable for the North Region and Roraima.

grew by 28 per cent but the leaseholders' conditions on these estates remained basically unchanged. Proprietor producers accounted for nearly 90 per cent of the area in 1985. Figures were not available in 1985 for the North Region, but in 1980 the Region had acquired an increasingly different profile from that of the national figures and proprietor producers took a much smaller share of the land.

In the early 1970s, Roraima's land-tenure structure still resembled the national rather than the regional situation. However, proprietorship fell drastically from 1970–75 when the share of all farmland in the hands of occupant producers rose from 5.5 to 80 per cent. This was the time of major road building programmes and immigration from other states. During this period, land under occupant status expanded faster than the total farmland area, suggesting that previously exploited tracts were being more intensively occupied. However, in the second half of the 1970s, following this wave of settlers, the area of land owned by proprietors grew whilst the proportion utilised by occupant producers, fell. This suggests that many areas formerly exploited by occupants came to be used by the proprietors, either as a result of titling or substitution.

The cycle of resource utilisation on land newly incorporated to rural estates is well known for Brazilian Amazonia, and its potential consequences are worrying for Roraima. Initially associated with road development, migrant influx and land appropriation, the production of seasonal crops and livestock predictably grew during the 1970s (Table 1.9). Cattle rearing remained the single most important source of income (except for the largely unknown mining production). If Roraima continues to follow the pattern of land development typical of the rest of Brazilian Amazonia, the increasing conversion of Roraima's forest land to pasture is likely to continue and is bound to exert pressure on forests, on the woodlands scattered across the savanna, and on the borders of Indian and ecological reserves.

An indication of this trend comes from changes in the pattern of farm size over the past decade or so (Table 1.10). In Brazil as a whole, the farm structure remained surprisingly stable over the period 1970–85. Until 1980, all size categories below 10,000 ha declined to the benefit of larger categories, but after 1980 the larger estates began to shrink (increasing the numbers in the 10 to 100 ha range). Even so, in 1985 farms with more than 10,000 ha managed to retain their 1975 level. In the North Region, the profile was initially more concentrated than for Brazil as a whole. It then followed the national trend towards subdivision into smaller holdings.

Changes were dramatic in Roraima. The profile was initially more concentrated in the categories over 1,000 ha than in either the regional or national figures. However, the profile of landholdings then underwent a more pronounced percentage redistribution towards smaller-sized categories than anywhere else. Despite this, newly incorporated land tended to be in the form of estates which were relatively large by regional and national standards (mostly 100–1,000 and 1,000–10,000 ha). During the more recent 1980–85 period, land incorporation proceeded more slowly everywhere. In Roraima, the net absolute reduction in total estate area was perhaps for a less durable and more damaging mixture of reasons than elsewhere in Amazonia. Exceptionally long dry seasons and local aggravation of seasonal water deficits in an area undergoing rapid population growth might be

Table 1.9 Total value and percentage distribution of livestock, crops and forest products, 1970–80

	All Total (Cr$)	Livestock (%)				Crops (%)			Forestry/ horticultural (%)	Native forest products (%)
		Total	Large	Medium	Small	Total	Perennial	Seasonal		
Brazil										
1970	24,967,914	32.20	21.16	5.39	5.65	59.67	11.65	48.02	1.56	6.57
1975	139,106,514	34.50	23.71	4.89	5.90	59.28	15.61	43.67	1.55	3.95
1980	1,542,298,299	39.35	27.17	4.68	7.50	55.23	12.46	42.77	2.37	3.05
North										
1970	729,693	18.55	11.76	1.66	5.13	55.43	13.04	42.39	0.00	26.01
1975	3,719,580	20.93	12.75	2.13	6.05	59.28	10.66	48.62	0.58	19.21
1980	58,985,956	23.42	15.62	2.42	5.37	51.70	13.50	38.21	1.53	23.35
Roraima										
1970	20,167	58.59	52.57	1.69	4.33	28.33	6.59	21.75	–	13.08
1975	76,338	58.45	43.27	3.10	12.08	38.05	4.11	33.94	0.38	3.12
1980	1,028,890	54.73	45.25	3.45	6.04	42.59	4.63	37.96	1.31	1.37

Sources: IBGE (1975) VIII Recenseamento Geral – 1970, Censo Agropecuário Brasil, Série Nacional vol. III, p. 175 (Rio de Janeiro: IBGE); IBGE (1979) Censos Econômicos de 1975, Censo Agropecuário Brasil, Série Nacional vol. i, p. 271 (Rio de Janeiro: IBGE); IBGE (1984) IX Recenseamento Geral – 1980, Censo Agropecuário Brasil, Série Nacional, vol. 2, tomo 3, no. 1, p. 291 (Rio de Janeiro: IBGE).

PERSPECTIVES

Table 1.10 Total area in rural estates and percentage distribution according to area categories of estates, 1970–85

	All	Area categories (hectares)				
	Total	Less than 10	10–100	100–1,000	1,000–10,000	More than 10,000
Brazil						
1970	293,012,168	3.11	20.53	37.17	27.44	11.75
1975	323,896,062	2.78	18.58	35.79	27.75	15.11
1980	369,587,872	2.43	17.44	34.35	28.59	17.19
1985	376,286,577	2.67	18.52	35.05	28.81	14.96
North						
1970	22,482,707	1.68	15.51	36.37	25.74	20.70
1975	32,615,964	1.67	13.47	26.38	20.11	38.36
1980	42,546,027	1.34	15.78	29.30	23.24	30.33
1985	44,884,354	1.38	20.63	30.21	21.08	26.70
Roraima						
1970	1,586,406	0.08	0.64	17.28	79.29	2.71
1975	1,836,201	0.25	0.93	9.67	51.23	37.93
1980	2,478,767	0.07	1.14	14.23	60.21	24.35
1985	2,157,128	0.05	6.73	22.69	57.34	13.18

Sources: IBGE (1974, 1979, 1984, 1988) *Anuários Estatísticos do Brasil 1973, 1978, 1983, 1987/8*, pp. 171–7; 377; and 312, respectively (Rio de Janeiro: IBGE).

thought to constrain the pace at which new land was incorporated. However, these very factors might also induce producers to intensify pressures on unexploited forest land on their holdings. This seems to have been the case in the 1980s.

The tendency for cattle ranches to encroach upon woodlands and continuous forests was already evident on maps derived from the Projeto Radam surveys in the 1970s. The land-use profile in Roraima is unusual by either Amazonian or Brazilian standards. During the 1970–80 period, as shown in Table 1.11, it was largely dominated by pasture. As has been shown earlier, Roraima possesses some of the very few natural grasslands in the North Region. More significantly however, the area under planted pasture which had been much less than the regional level, expanded greatly during the decade 1970–80. The increase in the planted area was remarkable, from less than 25,000 ha in 1975 to more than 82,000 ha in 1980, despite the low potential envisaged by the land evaluation survey.

Not surprisingly, and despite more than 340,000 ha having been incorporated into rural estates between 1970 and 1975, there was a net loss of 35,000 ha of forest and woodland over the same period. In 1970, these environments retained an importance comparable to the national profile but already lower than the regional average. During the 1970s, private appropriation in Roraima proceeded at a much faster pace than on the national or regional scene. This was particularly true in the late 1970s, following major road construction and migrant influxes. During the

33

Table 1.11 Total area in rural estates and percentage distribution according to land uses, Brazil, the North Region and Roraima, 1970–85

	All Total (ha)	Cropfields			Pastures			Woodlands		Not in use
		Total	Perennial	Season	Total	Natural	Planted	Total	Natural	Planted	
Brazil											
1970	294,145,466	11.55	2.71	8.84	52.40	42.29	10.11	19.68	19.11	0.56	11.36
1975	323,896,032	12.35	2.59	9.76	51.15	38.89	12.26	21.83	20.95	0.89	10.14
1980	364,854,421	13.46	2.87	10.59	47.83	31.22	16.61	24.17	22.79	1.37	9.16
1985*	376,286,577	13.92	2.61	11.31	–	–	–	–	–	–	–
North											
1970	23,182,145	2.66	0.57	2.09	19.10	16.35	2.75	60.07	59.83	0.20	14.73
1975	32,615,964	3.66	0.73	2.93	16.19	11.37	4.82	66.20	65.94	0.27	9.60
1980	41,559,420	4.20	1.29	2.91	18.58	9.51	9.07	63.15	62.67	0.47	9.74
1985*	44,884,354	4.50	1.49	3.01	–	–	–	–	–	–	–
Roraima											
1970	1,594,397	0.32	0.10	0.22	71.94	70.56	1.38	20.47	20.45	0.02	1.74
1975	1,836,201	1.48	0.50	0.98	73.69	72.19	1.50	15.86	15.86	–	2.00
1980	2,463,107	1.19	0.17	1.02	65.03	61.69	3.34	25.06	25.06	0.00	3.31
1985*	2,157,128	1.31	0.22	1.09	–	–	–	–	–	–	–

Sources: IBGE (1975) VIII Recenseamento Geral do Brasil – 1970 Cense Agropecuário Brasil, Série Nacional, vol. III, pp. 440–1 (Rio de Janeiro: IBGE); IBGE (1979) Censos Econômicos de 1975, Censo Agropecuário Brasil, Série Nacional, vol. 1, pp. 246–7 (Rio de Janeiro: IBGE); IBGE (1984) IX Recenseamento Geral – 1980, Censo Agropecuário Brasil, vol. 2, tome 3, no. 1, pp. 264–5 (Rio de Janeiro: IBGE); IBGE (1988) Anuário Estatístico do Brasil 1987/8, p. 311 (Rio de Janeiro: IBGE).
Note: * 1985 preliminary figures unavailable for areas under pastures, woodlands or unused.

1975–80 period, forested areas increased as a proportion of total estate land use from 16 per cent (291,185 ha) to 25 per cent (617,368 ha). Crops represented a minimal and always less important share than in the regional and national figures, but absolute and relative growth was markedly greater. Seasonal crops, typical of the small-scale subsistence front of settlement, multiplied nearly sevenfold, especially from 1970 to 1975, rising from 3,658 to 17,859 ha.

CONCLUSION

It appears that whilst forest lands are being disturbed at a pace and to an extent out of proportion to the range of potential uses that they offer, they still represent valuable capital for smallholders and larger landowners. This will be explored in later chapters. Such uses include indirect values such as preservation of regional climate patterns and ground water supplies as well as direct effects on biodiversity, runoff, soil conservation and nutrient cycling. Pressures are likely to increase as a result of natural expansion and government encouragement. Spontaneous increases in farming enterprises will probably rise as road connections to the south are improved, and if government plans for the first half of the 1990s are fulfilled.

Concern over forest land privatisation in Roraima is fed by what were accelerating trends in the early 1980s. These are likely to remain in the 1990s,

Table 1.12 Total personnel engaged and number of tractors, and land/workers and land/tractor ratios, in rural estates in Brazil, the North Region and Roraima, 1970–85

	Personnel engaged	Tractors	Land/worker hectares	Land/tractor hectares
Brazil				
1970	17,582,089	165,870	16.67	1,776.52
1975	20,345,692	323,113	15.92	1,002.43
1980	21,163,735	545,205	17.46	667.89
1985	23,273,517	652,049	16.17	577.08
North				
1970	934,024	1,127	24.07	19,949.16
1975	1,412,647	1,733	23.09	18,820.52
1980	1,781,611	6,295	23.88	6,758.70
1985	2,230,203	6,082	20.13	7,379.87
Roraima				
1970	8,277	5	191.79	317,481.20
1975	19,044	29	96.42	63,317.28
1980	16,903	127	146.65	19,517.85
1985	21,197	126	101.77	17,120.06

Sources: IBGE, (1984, 1988) Anuários Estatísticos do Brasil 1983, 1987/8, pp. 372 and 311, respectively. IBGE (1984) IX Recenseamento Geral 1980. Censos Agropecuários Brasil, vol 2, tome 3, vol. 1, p. 276 (Rio de Janeiro: IBGE).

accentuated to some extent by the profits from gold mining, and include rural estate mechanisation, credit allocation for cattle production and wood extraction.

From the early 1970s, road construction was a characteristic feature of life in Roraima, being markedly concentrated in the north around the urban nucleus of Boa Vista, with ranching, forest clearance and seasonal agriculture dominating the rural economy. Inward migration was stimulated by government colonisation projects and the gold-rush towards the end of the 1980s. By the late 1980s, Roraima was taking a greater and greater share of forest development within the North Region as a whole. The figure represented 2.4 per cent of the North Region from 1970–75 but had risen to 7.2 per cent by 1980. This was followed by a period during which large tracts of forest were appropriated with dramatic increases in planted pastures. The pace of uncontrolled private access to forest land was and is alarming.

During the period from 1970 to 1985, the area served by every farm worker and tractor diminished much more rapidly than at the regional level (Table 1.12). In fact Roraima reached mechanisation levels in 1985 that were comparable to those held in the North Region only ten years earlier. Additionally, a reversal in the allocation of rural credit was evident in 1985, which at that time clearly favoured large livestock undertakings (Table 1.13). This favours an accentuation of the trends indicated earlier. From 1980–85, credit assigned to livestock activities increased by over ten times that

Table 1.13 Total credits granted by Banco do Brasil to agricultural and livestock activities (1970, 1975), and funds granted by the national rural credit system to individuals and cooperatives (1980, 1985), and percentage distribution according to type of activity, in Brazil, the North Region and Roraima, 1970–85.

	All Total*	Agriculture (%)	Livestock (%)
Brazil			
1970	4,306,081	81.83	18.17
1975	53,623,385	76.65	23.35
1980	859,193,000	81.36	18.63
1985	51,705,203	91.57	8.43
North			
1970	41,351	73.76	26.24
1975	525,125	60.05	39.95
1980	26,083,000	88.27	11.73
1985	680,288	74.41	25.59
Roraima			
1970	1,167	12.51	87.49
1975	29,293	1.87	98.13
1980	794,000	76.07	23.93
1985	11,150	25.33	74.67

Sources: IBGE (1972, 1977, 1983, 1988) *Anuários Estatísticos do Brasil 1971, 1976, 1983, 1987/8*, pp. 468; 422; 391; and 326, respectively (Rio de Janeiro: IBGE).
Note: All values in Cr$ thousands. 1980 figures were published originally in Cr$ millions.

Table 1.14 Volume/weight of timber, charcoal and fuelwood produced from native forest species: Brazil, the North Region and Roraima, 1970–1985

	Timber m^3			Charcoal (tons)			Fuelwood (thousands of m^3)		
	Total	Change (%)		Total	Change (%)		Total	Change (%)	
Brazil									
1970*	–			1,589,556	–		134,804,033	–	
1975	31,527,909	–		2,396,237	50.75		122,069,682	−9.45	
1980	36,211,589	14.86		2,519,731	5.15		128,115,884	4.95	
1985	42,884,197	18.43		3,514,809	39.49		139,729,768	9.07	
		%NR/NR			%NR/BR			%NR/BR	
North									
1970*	–		–	13,963	–	0.88	3,249,446	–	2.41
1975	4,534,424	–	14.38	25,427	82.10	1.06	5,795,317	78.35	4.75
1980	11,483,489	153.25	31.71	30,827	21.24	1.22	8,307,827	43.35	6.48
1985	19,793,218	72.26	46.16	42,275	37.14	1.20	19,784,281	138.14	14.16
		%RR/NR			%RR/NR			%RR/NR	
Roraima									
1970*	–		–	33	–	0.24	–		–
1975	14,297	–	0.32	47	42.42	0.18	13,260	–	0.23
1980	72,857	409.60	0.63	28	−40.43	0.09	63,091	375.80	0.76
1985	39,920	−45.21	0.20	40	42.86	0.09	67,936	7.68	0.34

Sources: IBGE (1972, 1981, 1982, 1983, 1984, 1986, 1988) Anuários Estatísticos do Brasil: 1971, 1980, 1981, 1982, 1983, 1985, 1987/8, pp. 373; 344; 374; 430; 286; 364; and 350 respectively (Rio de Janeiro: IBGE).
Note: * 1970 Census figures on timber production unavailable.

apportioned to arable farming. A very approximate and certainly conservative indication of growth within forest areas is the increased production of basic wood staples. The fivefold increase in the amount of timber and the amount of fuelwood generated between 1975–80 coincides with the massive incorporation of forest land that occurred with pasture expansion. Possibly as a result of reductions in timber demand by crisis-stricken Venezuela and/ or export drives by Brazil in the early 1980s, the sustained growth of fuelwood is evidence of wastage of valuable forest resources under conditions of an insufficiently developed market (Table 1.14).

The state trends documented here through official census figures provide a general framework for the other field-based contributions in this book. Trends will not only be confirmed through the discussions of specific surveys but, in addition, an effort will be made to understand the rationale behind the decision-making processes which brought about these changes. Most of the farmers involved in the quickening pace of land-use change are not large landowners but they do live with the dream of one day becoming rich cattle ranchers. How may land development cope with such expectations in the face of long-term environmental and social problems? How can it offer reasonably attractive alternatives? Such high expectations are almost certainly likely to fail, given the known history of land exploitation elsewhere in the Amazon, so where can they now turn? The options are likely to be more limited than has been the case so far in Amazonia, if only because for most of them, Roraima, at the political frontier, is quite literally at the end of the road.

Figure 1.16 Gallery forests on *várzea* land in a savanna landscape, along the Rio Brava north of Boa Vista, showing sand bank exposed at low water in the dry season.

2

INDIANS, CATTLE AND SETTLERS

The growth of Roraima

John Hemming

INTRODUCTION

Sixteenth-century romantics such as Antonio de Berrío and Walter Raleigh imagined that the elusive and illusory kingdom of El Dorado might lie near the headwaters of the Essequibo or Branco; but it was not until the early eighteenth century that the first Europeans set foot in this region (Hemming 1978). From about 1720 onwards, a few Portuguese and Dutch slavers and slave traders were taken up both of these rivers to Roraima by Indian guides and paddlers. They returned with turtles, fish and native slaves that they had seized or bartered from the tribes of the upper Rio Branco.

The Treaty of Madrid in 1750 divided most of South America between the two Iberian kingdoms, Spain and Portugal. It sought to recognise exploration or occupation by Europeans, and therefore awarded the Branco Basin to Portugal on the strength of the handful of Portuguese adventurers who had ascended the river on their nefarious traffic. The Treaty of Madrid also chose geographical features to mark the boundary between the two colonial empires. It fixed it as the watershed between the Amazon and Orinoco, and this unambiguous line has survived without dispute until the present.

The provisions of the Treaty of Madrid were cancelled in 1761 and renewed in 1777, so that for sixteen years there was no treaty governing the frontier between Portuguese and Spanish South America. It was during that lull that Spanish missionaries and soldiers ventured south from Venezuela and established two fortified hamlets in what is now Brazilian Roraima. The Portuguese authorities on the Rio Negro reacted vigorously to this incursion. They sent a strong force north up the Branco. All the Spaniards in Roraima were soon rounded up; a fort was built to impose Portuguese rule and some thousand Indians were persuaded to settle in villages under colonial control. The native villages soon failed: their inmates rebelled and fled, and those that were recaptured were exiled to distant parts of Amazonia. However, during the 1780s, teams of brilliant Portuguese explorers ascended and mapped most of the rivers of Roraima.

It was also during the 1780s that the first cattle were taken up the Rio Branco to the grasslands of its headwaters. Ever since then, cattle have shaped Roraima's history. There were some 900 head by 1800 and the herd grew slowly and steadily throughout the nineteenth century. A few dozen animals were transported downriver each year on barges that took three months over the laborious round trip to Manaus. By 1900 there were around 60,000 head on two large, government-owned ranches and some forty private estates. Only a few hundred whites lived in the region, and there was only one small town – Boa Vista, located on the banks of the Rio Branco at the site of one of the eighteenth-century Indian villages.

This chapter traces the evolution of Roraima during the twentieth century. It deals in turn with the Indians, with colonising settlements and cattle, with demographic growth, and with Roraima's economic development.

THE FIRST OCCUPANTS: INDIANS AND THE INDIGENOUS POPULATION

Apart from a brief effort by the missionary Frei José dos Santos Inocentes in the 1840s, no one bothered with the welfare of the Indians of Roraima from the earliest contacts to the early twentieth century. This left them exposed to abuse and exploitation; but it also meant that there were no attempts to change their religion or culture. To some extent, tribes were able to decide for themselves whether to integrate and co-exist with the colonisers or whether to retreat farther from them.

The Arawak-speaking Wapixana adapted most easily to European influence (Figure 2.1). The Wapixana had always been exposed to this influence, mainly because they have traditionally lived very close to the main rivers Branco and Uraricoera. Because they tend to be submissive, according to Koch-Grünberg (1913), their prolonged relations with the white and mestizo population meant that they lost much of their own character. Stradelli commented that the Wapixana had retreated away from the rivers, that they were hard-working and docile and that they voluntarily worked for the colonists (Stradelli 1889: 264).

The Wapixana had been driven into three distinct regions by the advance of the Carib-speaking Makuxi. These areas were (i) on either side of the lower Uraricoera, Amajari and Parimé rivers and in the Taiano hills south of the lower Uraricoera, (ii) a few isolated groups amongst Makuxi on the Surumu–Cotingo in north-east Roraima, and (iii) the largest concentration in the Serra da Lua, south-east of Boa Vista, and eastwards into the upper Rupununi in Guyana (see also Figures 1.2 and 1.14). It was their misfortune that the Wapixana lived on some of the best cattle lands near Boa Vista. Ranchers regarded Indian land as 'unoccupied' and theirs for the taking, and they were able to enlist Wapixana men and women to work in their fields and houses for virtually no pay.

40

Figure 2.1 Linguistic groups of indigenous people in the 1970s.
Source: Modified from Migliazza 1978.

Coudreau contrasted the fate of the docile Wapixana with that of the more independent Makuxi and commented 'It is curious to note that tribes who become acculturated fastest also disappear quickest' (Coudreau 1886). The Wapixana became acculturated faster than the Makuxi and many of them learned Portuguese. As a result, the Wapixana, who were at one time the most important tribe of the Rio Branco, had diminished greatly by 1840 (Schomburgk 1840a). The Makuxi on the other hand, were far more numerous in the mid-nineteenth century than in the preceding hundred years and exceeded the Wapixana by two or three times, according to

Schomburgk. The Makuxi excelled as cowboys and the labour on the São Marcos national *fazenda* came exclusively from this tribe. Both Makuxi and Wapixana provided crews for cattle boats down the Rio Branco, despite the lingering animosity between them. Nineteenth-century travellers described both Wapixana and Makuxi as handsome, hospitable and good-natured. Many of these Indians continued to live in tribal villages far from colonial society but often suffering at the hands of white settlers. Slavery of Brazilian Indians had been forbidden by law since 1755, but continued under other guises. Nineteenth-century accounts of life in Amazonia by both Brazilian and foreign authors are full of descriptions of persecution of Indians, who tended to be gullible, subservient and uncomplaining.

Some violent settlers attacked Indians from the earliest phases of colonisation although most ranchers (*fazendeiros*) simply exploited the Indians. They evolved a 'godfather' system that has continued to the present. A rancher acts as a kindly paternalistic boss who protects his Indians and advances them goods on credit. He is godfather to their children and takes them into his own household, theoretically for education but in practice as unpaid domestic servants. Rivière (1972) found that, as late as 1967, there was extreme discrimination against these *filhos de criação* in some foster families. They were made to do all manual chores and to live and eat apart from the family's own children. Diniz (1972) called the system disguised slavery; but the Indians tolerated it, possibly because they were too weak and uncoordinated to resist. When the fostered Indian boys grew to manhood, they became cowhands receiving no pay beyond a few animals and some land to farm. Myers noted that

> this training has such deeply modifying effects on the Indian psyche that many of the individuals so brought up are more akin to the simple Brazilian ranchers of the region than to their own folk, speaking Portuguese much better than their own tongue and preferring the civilised way of life to that of the indian village (*maloca*).
>
> (Myers 1988: 19)

Myers noted that this acculturation was a two-way process and that Indian skills and labour made it possible for the colonists to establish themselves on these plains.

Indian population: its decline and resilience

A more serious threat to Roraima's Indians came from imported diseases against which they had no inherited immunity. Most of the tribes listed by eighteenth-century authors or by Schomburgk in 1840 had been extinguished by the early twentieth century, either by forced exile in the 1780s or more often by disease. Invasions of their lands by ranchers and forced labour caused social disruption, and this doubtless reduced the tribes' already low birth-rate.

In Schomburgk's day, the Sapará had a large village north of Maracá island on the Uraricoera, but Koch-Grünberg in 1911 could find only a few remnants of this tribe, dispersed among other Caribs. The Waimara were few and sickly when Schomburgk saw them on the upper Uraricoera; and Koch-Grünberg found them reduced to two brothers, who showed him the ruins of their former villages. The Yekuana (known to the Makuxi as Maiongong and to Venezuelans as Maquiritare) were great navigators who moved between the upper Orinoco and the Uraricoera. They were also sadly diminished from measles, malaria and work in rubber tapping in Venezuela. The Taurepang (known to the whites as Arecuna, and in Venezuela and Guyana as Pemon) were a Carib-speaking tribe closely related to the Makuxi, who lived on the Surumu and the slopes of Mount Roraima, and westwards to the tip of Maracá island. Koch-Grünberg said that the Taurepang had once been almost as numerous as the Makuxi, but were reduced to 1,000 or 1,500 in his day, from smallpox and other diseases.

The Makuxi were an exception to this pattern of demographic decline. Schomburgk (1840b) estimated that there were 3,000 Makuxi in 1839, of whom roughly half lived in Brazil and half in British Guiana. This tribe's numbers were reportedly similar in the first decade of the twentieth century.

Then the Makuxi were also devastated by disease. There was a particularly bad attack of measles in 1910 that killed thousands of Indians. This was followed in 1911 and 1912 by an unusually severe drought followed by wide-ranging fires in the lower Rio Branco and Rio Negro region. Koch-Grünberg wrote that many plains Indians died of starvation, since they had none of the plentiful game of forest Indians and depended on plantations. The savannas were said to look as though they had been burned and a large proportion of the cattle died (Dom Adalbert 1913). By November 1912 it was reported that a great epidemic of fever was raging, with the Indians dying *en masse*. The situation had not improved by 1919, when a company prospectus said that 'bilious fevers' had first struck Rio Branco in 1909, before which malaria had been unknown there. The disease arrived from the north, and recurred annually. It struck both colonists and Indians, and thousands died from it. The prospectus warned that if the authorities did not act quickly, Rio Branco would be a desert. Devastation from these diseases and drought reduced the Makuxi to 1,700 people by 1943, in both Brazil and British Guiana.

In the following year, Myers calculated that there was a total of 1,800 Makuxi and wrote about the concern at their diminution. She blamed malaria, but also noted that the infant mortality rate was high, as much as 50 per cent or more in the Canuku villages, and that respiratory diseases, chiefly bronchopneumonia, also took a large toll (Myers 1988). In the early 1940s, she also reported a serious outbreak of alastum smallpox.

Tribes other than the Makuxi were even fewer in numbers. The French engineer Maurice Mollard reckoned there were 3,000 Indians of all tribes in the region in 1913, and Dom Adalbert (1913) guessed 4–5,000 Indians in all Rio Branco in that same year. In the forests of the extreme north of Roraima lived a few hundred Ingarikó (known as Akawaio or Patamona in Guyana). These were feared as formidable warriors by the Taurepang and Makuxi, and lived beyond contact with the white frontier until recent decades. The large Yanomami tribe (formerly called Waika) and the related Xirianá also lived well beyond the colonial frontier, in the densely forested Parima hills between the upper Orinoco and the Uraricoera, Mucajaí and Catrimani rivers.

The boundary between Brazil and British Guiana (Guyana) was settled by international arbitration in 1903 but meant little to the Carib and Arawak-speaking tribes who had moved across the savannas since before the advent of colonists. The frontier does not lie on the watershed between the Branco–Amazon and the north-flowing Essequibo–Rupununi. That watershed is ill-defined, lying in seasonal lakes and open grasslands. At arbitration therefore, the Maú and upper Tacutu rivers were chosen as the frontier (see Figure 1.14), since they form a more obvious north–south line of demarcation.

Missionaries on either side of the frontier developed a rivalry to lure Indians, particularly the Makuxi. This rivalry was particularly strong from 1838 to 1842 when the English Protestant Reverend Thomas Youd and the Brazilian Catholic Father José dos Santos Inocentes clashed in their attempts to convert the Makuxi of Pirara village, which lay close to Lake Amucu on the watershed. The 'Pirara incident' led to armed occupations of Pirara by Brazilian and British troops, diplomatic protests, and the eventual neutralisation of the disputed territory until the arbitration of 1903.

Later in the nineteenth century, English missionaries, particularly Protestants of the Society for the Propagation of the Gospel, tried to convert the Makuxi and Wapixana of southern British Guiana. Using the lures of trade goods and dedicated teaching, they were quite successful, although large parts of both tribes preferred to live beyond any contact with either the British or the Brazilians, in the hills of northern and eastern Roraima. Several other missionary incursions and some rivalry between British Protestants and Brazilian Catholics occurred in the early decades of the twentieth century.

In 1910 the Brazilian government created the Indian Protection Service (SPI) to protect its native tribal peoples. Roraima's last surviving national *fazenda*, São Marcos, was awarded to the SPI to administer. For a time, using Makuxi cowhands and improved ranching techniques, São Marcos's administrators increased its herd. Gondim, in 1921, was impressed to find its cattle of good quality and numbering 8,000 – well up from the miserable 3,500 left by the last tenant rancher. The Makuxi were well housed and their children received primary schooling.

The Indian Protection Service sought to do more than merely manage its ranch, at this time. It opened more schools for Indians, a sanatorium on the Cotingo two days' ride from São Marcos, and an outpost called Limão on the upper Surumu for some Makuxi and Taurepang. When General Rondon, the great head of the SPI, visited the area and climbed Mount Roraima in 1927 he was able to visit these establishments. Curiously, however, SPI activity declined after Rondon's visit. It neglected its brief to protect Indians, and by 1944 Lima wrote that even the national *fazenda* São Marcos was 'in utter decay which is accentuated daily by its abandonment, without the introduction of any advanced techniques of cattle or horse breeding' (Lima 1944).

When General Rondon was on the upper Tacutu in 1927 a Makuxi chief complained that persecution by the local sheriff was forcing his people to seek greater freedom in British Guiana. Rondon commented: 'What a difference between the English of Guiana and the Brazilians on the frontier. The former seek to attract all the Indians of the region to their territory, the latter persecute their compatriots, forcing them into exile' (Rondon 1955). Another Brazilian general found the situation unchanged fifteen years later. He was impressed by the clever propaganda of British missionaries to attract Makuxi and Wapixana (Figueiredo 1944).

In recent years, Brazil has proved more attractive to the Indians than the independent republic of Guyana. This is partly due to Roraima's prosperity, which contrasts with the falling standard of living in Guyana. There has been a general antagonism between the Indian and black groups and hence to the government in Georgetown. This was accentuated by the disruptions following the short-lived attempt at rebellion or secession in the Rupununi in 1969.

Indians: survival of the fittest?

Roraima's Indian population divides into forest-dwelling tribes that have had little contact and savanna tribes that are now highly acculturated and almost assimilated into frontier society. The latter are the Makuxi, Wapixana and related Carib- and Arawak-speaking groups. The population of these semi-acculturated Indians has increased dramatically during recent decades, keeping pace with the growth of Roraima's total population.

In 1913 the native population of Rio Branco (Roraima) was estimated at between three and five thousand. That estimate was presumably only for savanna tribes, since there was then almost no contact with Yanomami or other forest tribes. In 1944 Myers reckoned that there were little more than a thousand Makuxi in Brazil (Myers 1945–46). In the next two decades, Ribeiro (1957) and Kietman (1966) guessed that there were 1,500–2,000 Makuxi, plus 1,000 or 1,500 related Taurepang and the same number of Wapixana in Brazil. Thus they reckoned 6,000 savanna Indians at most. By the 1970s, Diniz (1972) and Migliazza (1978) each calculated that Roraima's

Makuxi had doubled, to 3,000. But by 1986, the Indian population had possibly tripled in 35 years. Amodio and Pira, anthropologists with an intimate knowledge of the region's plains Indians, now estimate that the new state contains 12,000 Makuxi, 550 Taurepang, 5,000 Wapixana and 500 Ingarikó: a total of over 18,000 for these tribes. An authoritative and detailed census of Indian lands published in 1987 by the ecumenical research centre CEDI confirmed the population explosion. It listed a total of 15,395 people of these four tribes living in reserves (CEDI 1987; Amodio and Pira 1986; Hopper 1966; Diniz 1972; Migliazza 1978) – and it is known that many Indians now live outside tribal lands, particularly around Boa Vista.

There are various explanations for this increase. One may simply be that the modern counts are more accurate than earlier guesses. More importantly, Catholic and Protestant missionaries have been very active among these acculturated tribes, breaking down tribal customs, encouraging them to emulate settlers by having large families, and providing health and particularly natal care. Indians from these tribes also use the hospitals in Boa Vista and the rural health programmes provided by the government. After two centuries of contact, the Makuxi and Wapixana have developed some immunity to imported diseases that devastate newly contacted tribes. The Indian service FUNAI (which succeeded the discredited SPI in 1967) has increased its presence in Roraima and, together with the missionaries, has given some protection to native lands that were being engulfed by cattle ranching. This has improved the feeling of security of protected groups: their stability has resulted in population growth, from a higher birth-rate and lower mortality.

Indian culture and survival is closely linked to land. The plains Indians saw their territories ruthlessly invaded and usurped by cattlemen and farmers. They themselves had little understanding of legal ownership of land, which they regard as common to all mankind, and were powerless to resist. During the past decade, their constitutional right to their land has finally been recognised with the demarcation and protection of pockets of land around established *malocas*. The 1987 census of Indian lands listed 24 *malocas* occupying almost 400,000 hectares containing Makuxi and Wapixana. Eight of these *malocas* contain 1,244 Makuxi, 8 have 2,025 Wapixana, and 8 have 2,056 of both tribes mixed. In addition, there are several thousand members of both tribes among 9,186 Indians in the Ingarikó's vast (1,401,320 ha) Raposa/Serra do Sol reserve near Mount Roraima in the north-east of the state. The legal status of this large area is uncertain and it is under threat from ranchers and prospectors. There are also around a thousand Indians (Makuxi, Taurepang and Wapixana) in the old national *fazenda* of São Marcos, which still occupies 653,949 hectares despite severe depradations by a tenant rancher at the turn of the century. Pressure on Indian land is still intense. But the Indians now have vigorous legal defenders and they themselves have become more aware of their rights.

In recent years, Roraima's forest Indians have also been threatened. The Yanomami are the largest surviving tribe of forest Indians in South America. There are reported to be around 17,000 Yanomami, with roughly half in Venezuela and half in Brazil, living in hundreds of *malocas* scattered amid the densely forested hills of the watershed between the upper Orinoco and the Negro and Branco basins (Ramos and Taylor 1979; IBGE 1981). However, recent estimates suggest a much smaller number (see Chapter 7). Contact with the Yanomami was very rare until the mid-twentieth century.

The first call for the Brazilian territories of the Yanomami to be given the protection of reserve status was made by Ramos and Taylor in 1968. FUNAI and the Catholic Church joined their petition for a uniform park embracing all Yanomami lands, but in subsequent years FUNAI altered this to a request for seventeen isolated Indian areas – an archipelago of reserves surrounded by areas of colonisation or prospecting, which would be disastrous for the survival of this Indian nation. In 1979, Cláudia Andujar organised the Comissão pela Criação do Parque Indígena Ianomami (CCPY) that called for a park of 10 million hectares for the 8,500 Yanomami living in Brazil. An international campaign, led by Survival International in London and the American Anthropological Association, gave support.

Meanwhile, the threat to the isolation and cultural survival of the Yanomami came from two fronts. In the south-east of their territory, the BR 210 Perimetral Norte road was cut westwards from Caracaraí. Where it crossed the Catrimani river it came into damaging contact with some Yanomami: over a hundred Indians died in epidemics of measles and influenza brought by road workers and settlers. A more serious threat came from prospectors (*garimpeiros*). A myth arose throughout Roraima that the Yanomami lands (and in particular the Surucucus hills where there was a concentration of Indian *malocas*), were full of gold and cassiterite, mineral wealth in such quantity that it could pay off all of Brazil's foreign debt! From 1976 onwards, *garimpeiros* periodically invaded Yanomami land, despite occasional orders forbidding such invasions. During these years, the *garimpeiros'* cause was championed by two local deputies, Cavalcanti and Fagundes, who waged a determined campaign in the National Assembly in Brasília and in the press in Roraima. Fagundes declared:

> I intend to diminish the immense area of Roraima that is blocked for any economic activity. For example, I find it entirely just to leave outside the reserve the Apiaú *garimpo*, where there are 3,000 prospectors [actually 250] and no Indians for a radius of 150 kilometres.
>
> (Quoted in Hemming 1990a: 29)

On the opposite side, a meeting of Yanomami tribal leaders in 1984 wrote to the Indian Deputy Mário Juruna:

47

We, the Yanomami Indians, ask you to help us remove the prospectors from our Indian lands. The prospectors have for the past two years been invading Yanomami lands, extracting our gold, bringing diseases, coveting and taking our women, and pillaging our plantations.

(CEDI 1985)

Another Yanomami assembly in March 1986 reinforced the demand for the Yanomami Park to be properly demarcated and protected.

In March 1982, the Minister of the Interior signed a decree 'interdicting' 7.7 million hectares that would form a future reserve for the 8,400 Yanomami living in 92 *malocas* in Brazil. Of this vast area, some 5 million hectares were for the 7,100 Yanomami living in Roraima, and the remainder for those living north of the Rio Negro in Amazonas. Finally, after a determined campaign, in February 1992 the Yanomami territories were finally and fully designated.

In 1985, the Brazilian armed forces launched the Calha Norte campaign to plant military garrisons all along Brazil's northern frontiers (Allen 1992). Barracks were to be built and airstrips expanded at a series of strategic points, which were invariably Indian missions or FUNAI posts. The Calha Norte was never properly explained. It was said to be protection for this frontier; but there was no conceivable international military threat to this boundary that had been peaceful for over two centuries. It was reported to be intended to prevent drug smuggling or the entry of subversives – as though smugglers would try to enter Brazil across some of the toughest forests and rapid-infested rivers on earth. It was claimed that the presence of troops would bring the Indians into the mainstream of Brazilian society, a delicate process that was being achieved by missionaries and FUNAI officers.

The situation of the Yanomami changed dramatically from 1987–90, when a massive gold-rush brought tens of thousands of prospectors flooding into Roraima. The centre of the gold strike was around Paapiú, an abandoned airstrip built by the Air Force (FAB) as part of the Calha Norte programme. Paapiú is near the upper Mucajaí river, well within the interdicted Yanomami territory, some 40 km south-east of Surucucus. A team that inspected the area in June 1989 found the airstrip full of *garimpeiros*' planes, but no presence by any government agency. The Indians were suffering from malnutrition because their traditional game was frightened off by the miners' air traffic. Their rivers were poisoned by mercury used in the gold-panning process. The team reported that 'the Yanomami are suffering violent aggression to their culture by being exposed to uncontrolled and promiscuous contact with the *garimpeiros*'. Nothing was done, and by the end of that year many Yanomami died from an epidemic of virulent malaria unwittingly introduced by the miners. In 1990 the newly elected President Collor publically dynamited many clandestine airstrips; but it has proved extremely difficult to evict an estimated 40,000 poor and desperate *garimpeiros* excited by the lure of gold.

On the eastern frontier of Roraima, the Carib-speaking Wai-Wai tribe continues to live in relative isolation. Some of its members have come into contact with the eastern branch of the BR 210. Some 170 Wai-Wai live in Brazil, in a reserve of 330,000 hectares. In the south-east of Roraima, 418 Waimiri, Atroari, Piriutiti and Karafawyana live in a 2.44-million-hectare reserve, much of which is in the state of Amazonas. The Waimiri-Atroari were pacified in the 1970s after centuries of stubborn resistance, as part of the campaign to build the Manaus–Boa Vista road (see pp. 40–41) and suffered the inevitable depopulation from disease after contact.

CATTLE RANCHING

At the start of the twentieth century, Roraima had two huge government-owned national *fazendas*, which had been leased out to private tenants, and some forty other ranchers who raised cattle on the savannas that stretched away from the main rivers. Major Ourique (1906) wrote that by 1892 Rio Branco's cattle had increased to 60,000 with over 3,000 horses (Figure 2.2). This growth occurred 'without the slightest care, for breeding was left to the laws of nature, to the voracity of jaguars (which used to be very abundant in those parts) and to all manner of other accidents that no one tried to avoid' (Ourique 1906; Guerra 1957).

The rapid growth in the size of the cattle herd was largely the result of private ranchers taking over state lands. Intruders invaded and occupied the lands of the national *fazenda* of São Bento, which was divided into 'many prosperous private ranches on which are raised, with lively results, thousands of head of cattle, horses and sheep' (Ourique 1906). In the years before 1921, according to Gondim, 'some of these intruders alleged that the lands were unoccupied (*devolutas*) and demanded and obtained from various governors the issue of definitive title deeds. The two national properties thus disappeared from public ownership' (Gondim 1922). The one surviving national *fazenda*, São Marcos, occupied the 'V' between the Uraricoera and Tacutu rivers, bounded to the west by the Parimé and east by the Surumu, and it extended northwards towards the Pacaraima hills. This ranch was transferred between government ministries and eventually in 1916 passed to the new Indian Protection Service (SPI) to administer on behalf of the Ministry of Agriculture. Before that, its lands and cattle had been plundered by one of the tenants. The SPI's administrator of São Marcos complained bitterly that when his lease expired, the tenant rancher withdrew

with a herd of 20,000 head and title as owner of an usurped ranch called Flechal – which was nothing more than a good part of the

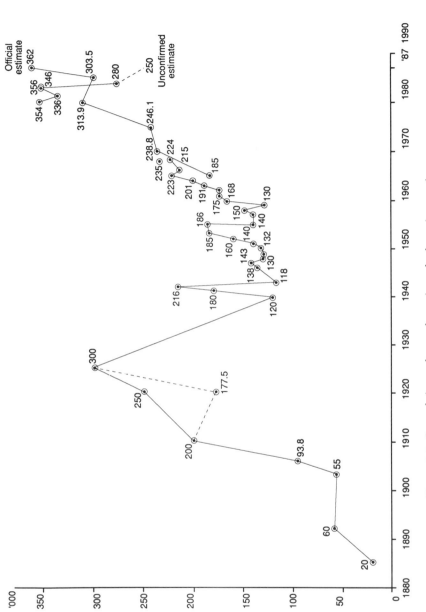

Figure 2.2 Growth in cattle numbers since the late nineteenth century.

national fazenda São Marcos. He left for the nation a miserable herd of little more than 3,000 cattle, all old cows and useless for breeding.

(Hemming 1990a)

The tenant argued that he had returned the 3,000 government-owned cattle of which he had taken custody in his original contract of 1888.

When this most active of entrepreneurial tenants died, his family inherited the Flechal Victoria *fazenda* and, following a legal action, won not only the usurped ranch but also damages from the state. A visitor said in 1911 that the Brazilian government had caved in because it feared revolution if it tried to control the proud *fazendeiros* of the isolated Rio Branco. The national *fazenda* São Marcos was vast – 8,000 square kilometres – and contained 18–20,000 head of cattle. However, only 5,000 of these definitely belonged to the government: the rest were branded with the mark of the late tenant. A large and aggressive Manaus rubber-trading company J.G. Araújo e Companhia Ltda. acquired the property in 1918. Litigation over the national *fazendas* and their cattle continued into the higher courts; but the state never recovered the ranch of São Bento or large parts of São Marcos, and the behaviour of the firm was indignantly attacked in the Annual Reports of the SPI. By the 1920s, J.G. Araújo was the largest cattle rancher in the area, with 45,000 head on a series of ranches, particularly in the bend of the river bank of the Uraricoera–Branco river north of Boa Vista. Across the river and up the Parimé were the lands of a further large-scale rancher with 13,000 head. Boa Vista's leading businessman at that time was Homero Cruz, who administered the J.G. Araújo herds as well as 5,000 head of other land-owners and his own 5,000 cattle on the Surumu. The Magalhães family, descended from Boa Vista's first rancher, had 5,000 head on two ranches north of the town (Hamilton Rice 1928).

The herds of Roraima grew steadily during the first decades of the twentieth century (Figure 2.2). A cattle census by the Ministry of Agriculture in 1912 gave 200,000 cattle and 6,800 horses. Another census estimate in 1920 gave a lower figure of 177,500; but some authorities reckoned 250,000 cattle and perhaps 80,000 horses for that same year. By 1925 it was reported that the herd had increased to over 300,000 head. It was generally agreed that the early 1930s were the golden age of cattle in Roraima, with well over 300,000 head; although all these figures, even those of the official censuses, were estimates or guesses. Disaster struck in the mid-1930s and, within a decade, the cattle herd was halved. In 1940, there were only 120,247 cattle and 12,073 horses reported. Local cattlemen gave a number of reasons for this terrible decline, including disease (especially rabies), impoverishment of the pastures, primitive breeding methods, and an exodus of labour for mineral prospecting. Roraima's cattle had been stricken by rabies as long ago as 1918. Rabies returned in 1931 in a far more virulent attack. According to Rivière (1972), the disease was invariably fatal and 'reached epidemic proportions, with some ranchers losing as much as half their stock'. Two varieties of cattle rabies were identified. The most common form was a paralysis that affected an animal's digestion and appetite, so that its grazing became irregular, after which its hind quarters became paralysed and it had to be destroyed. In a rarer 'furious' form, cattle became uncontrollably aggressive (Guerra 1957).

Throughout these years, the grasslands were burnt at least annually, at random intervals. This practice was borrowed from the Indians, who used burning as a method of hunting before the introduction of cattle. It was thought that this burning reduced pests such as ticks or rattlesnakes and it checked weeds and rampant growth of grasses. Farmers still burn the plains to this day, liking clear open plains from which all bushes have been removed by fire. They call such land 'worked over' (*lavrado*). However, some experts claim that this practice progressively degrades the plain's natural condition. The grass becomes increasingly coarsened and more widely spaced; native legumes disappear and the caimbé (*Curatella americana*, a typical savanna tree) become more gnarled (IBGE 1981). The soils of Roraima are naturally low in nutrients, and these deficiencies are heightened by annual burning. Such clearance of vegetation also exposes the soil surface to erosion. There was no attempt to improve the quality of the cattle by selective breeding. The herds simply roamed free. By the 1920s only a few ranches had fenced off sections (*retiros*) where cows could be placed after they gave birth. The plains were regarded as common land. Cattle grazed at will, often sheltering in the shade of palm trees by day and grazing on cooler moonlit nights.

For younger generations of Brazilians, Roraima has always been seen as a cattle frontier. However, the creation of the Federal territory of Rio Branco in 1943 came just after the terrible decade of the 1930s when rabies and foot-and-mouth disease had halved the size of the region's cattle herd to some 120,000 head. An observer reported that Rio Branco's cattle ranching in 1943 was 'simplistic and rudimentary. The plains are master of the cattle. The cowboys are onlookers who see little; they are interested only in how many animals die. Few ranchers have wire fences' (Paixão e Silva 1943). A Bank of Brazil report a few years later said that 'the cattle there is of small stature and because it is bred freely with no form of technical assistance, is subject to all sorts of disease which inexorably decimate it' (Banco do Brasil, in Guerra 1957). The result was small animals with little meat in which 'the skeleton weighs more than the muscles' (Gusmão n.d.). One of the federal territory's first governors described its breeding methods as 'the most rudimentary known' (da Costa 1949).

Twenty-five years later, little had changed. A research team from the Fundação Delmiro Gouveia in Rio de Janeiro made a gloomy report on the cattle of Roraima. The animals were by then various cross-breeds of humped zebu, such as nelore, gir and guzerat, which had been imported because of their greater resistance to tropical heat and ticks. Despite this,

> it is sad and apathetic cattle – even the breeding stock imported from other parts of the country which, . . . abandoned to its fate, rapidly wastes away. . . . The cattle get no supplementary food, so necessary in a region of poor pastures.
>
> (Ribeiro 1969)

The result was animals with weak bone structure and little meat, which averaged only 220 kg when ready for slaughter at three years.

Disease continued to ravage the herd. Kelsey reported, in the late 1960s, that rabies kept reappearing to decimate Roraima's cattle. There had also been 'countless frustrating attempts to eliminate aftosa (foot-and-mouth disease)' (Kelsey 1972: 144). Foot-and-mouth killed only about 5 per cent of the region's cattle, but it caused widespread loss of weight, and secondary infections and abortions among the surviving animals. The local *fazendeiro* was fatalistic about it. 'By and large he folds his arms, assumes a philosophical attitude and allows his cattle to die one by one to the point where periodically many *fazendeiros* are completely wiped out' (Kelsey 1972; see also Rivière 1972). Even though the government offered free veterinary service, the cost of buying and applying vaccination was thought to be too expensive. Thus 'these two diseases continue to be endemic, largely unchecked and unabated, inflicting tremendous losses' (Kelsey 1972).

In 1970 the largest cattle company was still J.G. de Araújo (now renamed Gado da Amazônia SA) which had entered the region over fifty years earlier. This company had some 28,500 animals on 24 ranches, but its management was very weak. The ultimate owner, Gomes de Araújo, never once visited his vast landholdings in Roraima. As with most large ranches, day-to-day management was left entirely in the hands of overseers (*capatazes*) and there was minimal investment. 'His ranches are characterized by very low levels of efficiency and productivity per unit of area, but his holdings are so vast and his inputs so small that he receives a large and easy income regardless' (Kelsey 1972). At that time, IBGE's Statistical Yearbook recorded over 2,000 ranches in Roraima, with a total land area of over 2.1 million hectares. However, over 1.5 million hectares of this (71 per cent) was owned by the 704 large ranches in the size range 1,000–10,000 hectares and three *fazendas* were larger than 10,000 hectares (see Chapter 1).

In recent years, the husbandry of Roraima's cattle has finally started to improve and modernise, thanks partly to government agencies such as ASTER and EMBRAPA. New breeds of cattle were brought up the BR 174 Manaus–Boa Vista highway. Most farms now have machinery for cutting forage. The animals are given salt and, for the zebu breeds, mineral salts and sulphur. Experiments have also been carried out with new types of grass, such as the tall *colonião* or imported *quicuio de Amazônia*. Fenced grass meadows (*capineiras*) are created near stands of well-watered buriti palms (*Mauritia flexuosa*) and milking is done in these enclosures for a year, so that the manured ground yields food crops or particularly rich pasture. Boa Vista has a modern abbatoir with four slaughterings a week, each of at least 60 animals. Thus, in the course of a year, some 12,000 animals are slaughtered, usually when they have reached the optimum age of 4 to 5 years. Despite improved breeding and husbandry, Roraima's cattle are still poor by Brazilian standards. They average 315 kg when slaughtered, which is not far above the Ministry of Agriculture's minimum weight of 255 kg, and their carcasses account for 45 per cent of this total. A number of animals (around 7,000) used to be exported to Manaus, many by river during the rainy season, taking over 2 days from Caracaraí. Since 1981, however, Boa Vista has been forced to import beef since increased production is not keeping up with increasing demand. The majority of Roraima's ranches continue to be near its capital Boa Vista, in the bend of the Uraricoera–Branco rivers. In recent years, however, there has been expansion of ranching along the BR 401 towards Guyana, on the Surumu river and around the cattle town Normandia. There are also new ranches on the plains towards Venezuela and in the Serra da Lua (former Wapixana homelands) south-east of Boa Vista (Figure 1.2). Farther south, cattle have been introduced into land cleared of forest along the western branch of the BR 210 Perimetral Norte, and many small ranches have sprung up along the BR 174.

POLITICAL TRANSFORMATION AND SETTLEMENT

Settlement

The region's first census was taken in 1883. The non-Indian settlers then totalled 384, a figure that included nine Venezuelans and six Portuguese. There were no slaves. When Coudreau went up the Rio Branco in 1884 he wrote that it was difficult to imagine how empty the lower river was. Along 500 kilometres of river banks he saw only six hamlets of Carib-speaking Paushiana – perhaps 250 Indians in all. Stradelli in 1887 described the lower river as a desert in which there were only four tiny hamlets. On the plains above the rapids only two or three hundred *civilisados* or non-Indians were reported (Stradelli 1889).

The situation changed gradually during the next quarter century. In 1902 the government of Amazonas state commissioned a census of every person living along the Tacutu river, as part of the Brazilian evidence of the boundary arbitration of the frontier with British Guiana. This census named 507 people, of whom 320 were Makuxi Indians, 53 were Wapixana or related Aturiaú, 116 were Brazilian settlers, and 8 English (Tenreiro Aranha, in Nabuco 1903). During the twenty years from 1886 to 1906 the number of ranches in the entire region grew from 80 to 142; but most of these belonged to settlers who were not resident. Very few non-Indians lived on the land. By 1913 the French engineer Mollard reckoned that 1,000 whites and 3,000 Indians lived in the open country of the upper river.

At the height of the rubber boom, at the beginning of this century, there was a brief revival of activity on the lower Rio Branco. For example, Mollard (1913) said that the lower river contained some fifty rubber-gathering posts, including five that could be considered as villages. After the rubber boom collapsed, the lower river was again very thinly inhabited (Pereira 1917). Four years later, the only activity on the hundreds of kilometres of forested river was extraction of balata rubber and Brazil nuts and some tobacco growing (Gondim 1922).

Political evolution

During the first four decades of the twentieth century there was minimal government intervention in Roraima (or Rio Branco as it was then known). The state of Amazonas exercised little control over its isolated northernmost region. The few local cattle barons were allowed almost total political freedom (Hemming 1990b).

The area was equally abandoned in religious terms. During most of the nineteenth century, Boa Vista received only occasional visits from a parish priest from the Rio Negro, who came every few years to perform marriages and baptisms. In 1909 some Belgian Benedictines went up the river and their leader had the title of Bishop of Rio Branco. He and the other foreign missionaries soon fell foul of the most powerful cattle barons, who had set up a masonic lodge. The authorities in Manaus tended to side with the influential Brazilian rancher against the outraged foreign clerics.

Over subsequent years, we hear of occasional visits to the area by government geologists or veterinary doctors and agronomists sent to try to improve cattle-ranching techniques. In 1927, the great champion of the Indians, General Rondon, made a brief inspection of Rio Branco's northern frontier. During the 1930s there were more thorough surveys of the borders with Venezuela and British Guiana led by Brás Dias de Aguiar. The Federal Indian Protection Service had a burst of activity in the 1920s and then lapsed into inaction. With the collapse of the rubber boom, the state of Amazonas suffered an economic decline. The result was less demand for Rio Branco's beef in Manaus, and an acute shortage of funds for any public works. The only attempt to stimulate Rio Branco's economy was a private venture by the German Bishop Eggerath. His grandiose scheme to invest in the region's infrastructure collapsed ignominiously when one of his relatives absconded with the monastic funds for his new company.

Creation of a federal territory

As early as 1933 a commission of the Sociedade de Geografia do Rio de Janeiro had suggested the creation of ten federal territories in frontier regions of Brazil. A decade later, the idea was implemented. President Vargas's *Estado Novo* saw frontier-land colonisation as a safety valve for rural unrest. Social confrontation could be forestalled without offending the major farmers, by settling landless poor on unclaimed frontier land. In 1943, the Federal Government claimed national security as its justification for carving five federal territories out of their respective states. It argued that it alone had the resources to stimulate colonisation of these remote areas and it assumed the legal right to create these territories because they lay on national frontiers (Marshall 1988).

On 13 September 1943, Federal Decree-Law 5812 created the federal territory of Rio Branco out of the state of Amazonas. The new territory included the entire basin of the Rio Branco, with extensions to adjacent tributaries of the Rio Negro to the west and east of its mouth. It has an area of 230,104 square kilometres, and has 958 km of international frontier with Venezuela and 964 km with Guyana. The territory's name was changed to Roraima, after nineteen years, by Law 4182 of 13 December 1962. It was felt that there were too many Rio Brancos in Brazil and a group of local politicians therefore lobbied for change. The local newspaper canvassed its readers on the name they preferred and they chose that of the 2,772-metre table mountain Roraima, at the northern extremity of the territory. The territory received statehood in 1988.

Population

An overview of the population of Roraima compared to the North Region and the national figures has been outlined in Chapter 1. This section concentrates on the special features characterising the demography of the state. Between 1940 and 1992 the population of the territory grew eighteen-fold, from 12,200 to 215,800 (Figure 2.3), but still at the end of this period had one of the lowest densities of population of any state. Despite this sparseness, Roraima's population has grown proportionally far faster than that of the country as a whole.

This rapid growth has come partly from a high birth-rate and partly from immigration. Roraima has had a youthful population. In 1960 a remarkable 38 per cent of its population was aged under ten – a higher proportion than in any other part of northern Brazil. In 1970, 34.5 per cent was under ten and 60.8 per cent under twenty (IBGE 1976, 1986; Zimmerman 1973). There has also been steady in-migration from other parts of Brazil. In the territory's first census in 1950 shortly after its separation from the state of Amazonas, over three-quarters of the population said that they had come

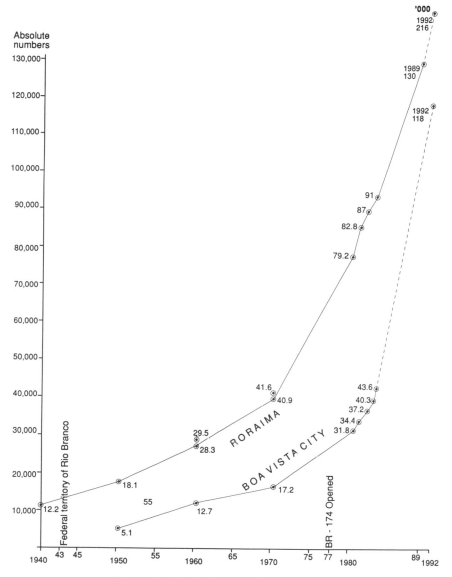

Figure 2.3 Population growth since 1940.

from elsewhere in Brazil. In later censuses, the ratio of non-natives fell but remains very high.

This inward movement shows up to some extent in statistics of passenger traffic. In 1976, when access to Roraima was possible only by air or river, there were 23,000 passenger arrivals by air. By 1978, after the opening of the road link, air traffic had fallen to some 16,000; but it rose steadily to over 34,000 in 1983 and to over 50,000 in 1984. The great increase associated with the gold-rush in the late 1980s has now subsided with tougher government action and recession, current estimates being rather less than the mid-1980s's figure. Meanwhile, of course, much traffic switched to the far cheaper buses on the new Manaus–Boa Vista highway (Roraima, SEPLAC 1980).

Buses cover the 775 km from Manaus in under thirty hours if all goes well and the 195 km from Boa Vista to Santa Elena de Guairén on the Venezuelan frontier in seven hours. These times will improve dramatically with the paving of more stretches of the road. Bus passenger traffic from Manaus rose from 14,000 in 1980 to 17,600 in 1984; and international traffic along the dusty 'Panamerican' highways to Venezuela and Guyana rose from 4,000 to 5,100 during the same period.

Government action and other external events have sometimes spurred inward migration. There have been military movements: with the arrival of road-building army engineers of the 6th BEC (Batalhão de Engenharia de Construção) in 1969, of a mechanised cavalry unit a few years later, and in 1987 when soldiers occupied frontier Indian posts in the controversial Calha Norte exercise. There have been planned migrations of settler families into agricultural colonies as shown elsewhere in this volume. There have also been periodic population booms caused by prospectors when word has spread of discoveries of gold or diamonds; although rushes are periodic, mining is also a historically significant factor in attracting migrants.

Boa Vista

Throughout the nineteenth century, the symbol of Brazilian rule of its northernmost territory was the small stone fort of São Joaquim at the junction of the Uraricoera and Tacutu rivers. However, from the middle of the century, a small village called Boa Vista developed a few miles downriver on the opposite bank of the Rio Branco. At the beginning of this century, Boa Vista was designated as a town and capital of a vast region (*comarca*) called Rio Branco.

Boa Vista was the only social centre for the cattle barons, but it was a sorry little place. Joaquim Gondim was disappointed when he reached the town in 1920. From a distance, its magnificent location, facing the Rio Branco and the outcrops of the Serra Grande, looked promising but closer inspection revealed small houses and wide potholed streets. Boa Vista had only 621 people in 1920 in 113 houses, some of which were built of masonry with tiled or corrugated-iron roofs, but most were of adobe or mud-and-lathe and only one shop possessed a water pump and an electrical generator. By 1924, Boa Vista's population had grown to 1,200 people who were 'hospitable and cheerful' according to a visiting geologist. This impression was in sharp contrast to that of the English novelist Evelyn Waugh, who was there a few years later (1934) and hated the place (Oliveira 1929; Waugh 1934).

Rio Branco's new political status in 1943 brought an influx of civil servants to run the territory and administer various development programmes. The town of Boa Vista was elevated to the rank of city and capital of the new federal territory.

The first governor sent from Rio de Janeiro, the energetic Captain Garcez dos Reis, took the bold and imaginative step in 1945 of employing the town-planning consultants Darcy A. Derenusson to design a plan for Boa Vista worthy of a capital city. Derenusson created a large civic centre, with the city's post office, hotel, main bank, Governor's Residence, telecommunications building and 'Palace of Culture' (now the state Legislature), all arranged around a broad, leafy square. The old riverside town became a shopping centre, with the civic centre at its landward edge. The new city radiates from the central hub in a spider's web of wide avenues (see p. 80), spreading across the flat savanna towards the airport and highways to Caracaraí and Venezuela. Although the new plan 'turns its back' on the Rio Branco, it has been a success. Boa Vista is one of Brazil's most agreeable small cities, an uncrowded place of mostly single-storey houses amid plenty of greenery. It has room to expand, although the built-up area has now surpassed the airport.

The construction of Boa Vista on Derenusson's plan was inevitably a slow process. By 1948 the territory's government was still housed temporarily in the bishopric while its future offices were being built. In that year, there were still only 22 passenger vehicles and 58 goods vehicles licensed in all Rio Branco, compared to over 500 horse-drawn carts and carriages. In 1952 Caldas de Magalhães complained that the partly built new city was a chaotic muddle (Caldas de Magalhães, in Guerra 1957).

The population of Boa Vista increased roughly in line with that of Roraima as a whole: from 5,132 in 1950 to some 118,000 in 1992 (Ferreira 1957; IBGE 1976; IBGE 1992). After the initial influx of civil servants and construction workers to build Boa Vista in the decades since 1945, much of the city's growth was organic. It had a young population and a slight preponderance of women to men. Planners were relieved to find that Boa Vista did not prove too powerful a magnet drawing people in from rural areas, although the Indian service FUNAI reckons that, in the late 1980s, there were as many as 15,000 Makuxi and Wapixana Indians living in and near the city, and that it does exert a strong urban pull. Boa Vista's infrastructure kept pace with its growth. Its airstrip was changed to an airport in 1958 and this was lengthened to take jets in 1972. The city acquired a football stadium, a waterworks producing some of the purest water of any South American city, a cathedral consecrated in 1972, the 1,200-metre steel-and-concrete 'Makuxi Bridge' over the Rio Branco opened in 1975, and a new bus service (badly needed by pedestrians in a city of broad streets and consistently high temperature). In the late 1980s there were plans for hydroelectric power from the controversial Paredão dam on the Mucajaí river, but energy is currently supplied by diesel generators.

ECONOMIC DEVELOPMENT AND AGRICULTURAL COLONISATION

Ever since the first European settlement of Roraima in the eighteenth century there have been official desires to stimulate agriculture – as opposed to cattle ranching – in the region. It seemed absurd that such an apparently fertile area should have to import basic foods and vegetables. It has been said that the *fazendeiros* (estate owners) were too conservative or idle to grow crops – their interest was only in cattle; but some subsistence crops were grown as there was (and at times still is) no reliable market or transport.

The demand for locally grown food accelerated with the creation of the territory of Rio Branco (later Roraima) in 1943. The newly arrived government officials and the growing population of Boa Vista wanted a diet more varied than meat, manioc and rice. The problem came to a head between November 1951 and March 1952 when the food shortage became so acute that the Governor had to appeal to the Brazilian Air Force for an airlift of rice, manioc, corn, beans and sugar.

The official response was a series of planned agricultural colonies. These had a triple objective: to make the region self-sufficient in basic foods; to attract immigrants to a very thinly settled part of Brazil; and to open cleared forest, as opposed to savanna, for farming. After a shaky start, the policy has been reasonably successful depending upon the point of view taken. In terms of political will to settle new land and implant agricultural settlements there is a measure of success, but in terms of sustainable and reasonable livelihoods or of supplying urban markets, the success is more arguable.

The first two colonies were founded in 1944, immediately after the creation of the territory: Fernando Costa on the banks of the Mucajaí, but later moved closer to Caracaraí, 76 km south of Boa Vista on the main road; and Brás Dias de Aguiar, only 30 km from Boa Vista, but south-east on the far bank of the Rio Branco and initially with bad communications. The first batch of settler families in these new colonies came from the north-eastern state of Maranhão. Each family got its passage, one month's adaptation in Boa Vista, some clothing and household utensils, a basic wage for ten months, some tools and seeds, and medicine when necessary. A 25-hectare plot was awarded to each family.

Despite this official help, the first pioneers soon left, as did most of another group of immigrants during 1947–48. Between 1951 and 1953, 140 families came to Fernando Costa and by the end of the decade it had 650 people and its own food-processing equipment. In those early days, farming methods were the very basic slash and burn – clearing forest and under-growth with machete, hoe and fire, and moving on to clear more land when the soil was exhausted, since there was no fertiliser. By the late 1950s, Brás Dias de Aguiar had grown more slowly because of the difficulty of moving its produce. It had only 58 families, growing mainly rice, but still opening fresh plots. Meanwhile, a third *colónia*, Coronel Mota (named after Roraima's first schoolmaster) was started in 1955 in the Taiano hills 92 km north-west of Boa Vista. The first settlers at Colónia Coronel Mota were also from the north-east of Brazil. In the following year, eleven Japanese families came, with support from their government. These tried to grow pepper and vegetables, but they were not trained farmers and soon moved away to southern Brazil.

Experience from these first three agricultural colonies showed that road access to Boa Vista was all-important, since the city was the only market for produce. Other problems stemmed from lack of technical assistance, shortage of rural credit and difficulty of storing food in the equatorial climate. Little mechanical help was available and soils were frequently prepared manually with hoes. Irrigation was good, with no shortage of water, and all the colonies pumped water from wells or streams. But there was no chemical fertiliser, only animal, and no preventive pest control beyond spraying to kill parasites as they occurred. Settlers in the early colonies had such difficulty getting their surplus produce to market that they often resorted to subsistence farming to feed their families, with no income to buy other essentials. A second generation of agricultural colonies learned from the trials of the colonies of the 1940s and 1950s, with plenty of advice and assistance from government agencies such as ASTER, EMBRAPA, ACAR-Roraima and INCRA.

In the 1970s a new colony was started at Alto Alegre on the upper Mucajaí, 100 km west of Boa Vista, with 92 families from Maranhão and advised by an advanced campus of the University of Rio Grande do Sul (see Chapter 5). Sorocaima is a later colony, formed spontaneously without offical planning on higher ground at 600 metres near the Venezuelan frontier. It is doing well, growing Colombian coffee. Prata was another planned colony, beyond Fernando Costa on the road to Caracaraí, 85 km south of Boa Vista. It was formed in the late 1970s with families from Amazonas and the north-east, and also some from Fernando Costa whose original soils were exhausted.

More recently, other agricultural colonies have been started, some spon-taneously, others planned (Figure 1.14). The opening of new highways in southern Roraima has enabled the government to plant new colonies along the BR 210 (Perimetral Norte) and BR 174. One problem has been a tendency to clear forest along these new roads in order to make pasture for cattle rather than for the production of crops. The few large ranchers employ gangs of itinerant labourers to fell rain forest and forbid them to plant any crops for fear that they might acquire settlers' rights if they did. These contract tree-fellers 'move-on, leaving great areas of burned land behind them. The natural resources are thus being squandered by bad use of the soil' (IBGE 1983; Guerra 1957; Ribeiro 1969; Zimmermann 1973; SEMTUR 1986).

Mineral prospecting

The monumental statue in the middle of the square that forms Boa Vista's civic centre is of a *garimpeiro*, and this symbolises a historically important contribution to life in Roraima. *Garimpeiros* are migratory adventurers, working in small teams of men without families, who push up rivers to prospect wherever there is a rumour of gold or diamonds. Although it may be argued that they have made little lasting contribution to the settlement or economy, they sometimes made a dramatic impact on the environment and some of the effects of successful strikes have been fed back into the process of land development, as is demonstrated in Chapter 7.

At times in Roraima's recent history, gold and diamonds have attracted waves of immigrants from other parts of Brazil or have lured labourers away from ranches and farms. But as soon as a river's yield declines, the prospectors move on, often across frontiers into Venezuela or Guyana. Their prospecting produces a steady supply of diamonds or gold dust, but both commodities are of high value and easily smuggled, so that little return from prospecting enters Roraima's official taxable economy. The *garimpeiros'* primitive methods of pumping and panning destroy river beds and probably miss a considerable proportion of the potential mineral wealth.

As early as 1920, Gondim spoke of a small number of prospectors seeking mineral riches on remote rivers. There was a diamond- and gold-rush in 1940, a year of agricultural drought and the lowest ebb of Roraima's cattle herd. The 1940s were a peak period for prospecting, with discoveries of some 11,000 carats a year, but the mother lode of diamonds has never been discovered, so that each area of *garimpagem* is temporary. In the 1940s, prospectors followed Brás Dias de Aguiar's boundary commission into the upper Amajari and the area around Mount Tepequém (Figure 1.14). In 1960 *garimpeiros* started to invade the upper Cotingo in the extreme north of Roraima, but they were prevented from entering its upper forests by the warlike Ingarikó. The northern rivers, such as the Suapí and other tributaries of the Quinô and Maú, and recently the Uraricaá and tributaries (such as the Ericó) have been active areas for prospecting. Seven per cent of Roraima's population was reckoned to live from *garimpagem* and its annual yield was thought to be over 80,000 carats of diamonds and 750 kilos of gold (Gondim 1922; de Aguiar 1942; Ribeiro 1969).

A decline in prospecting activity was dramatically reversed in 1987, when gold was discovered on the upper Mucajaí – within the limits of the

Yanomami Indian park. This discovery brought a massive gold-rush flooding in from all parts of Brazil, and provoked both national and international attention, leading to a final settlement of the Yanomami lands question early in 1992. At its peak, this gold-rush attracted up to 40,000 prospectors and their uncontrolled clandestine flights made Boa Vista the busiest airport in South America. They moved on only when the finds of gold declined.

The role of transport in development

The most serious constraint on the growth of Roraima's cattle industry and therefore of the territory, was difficulty of transport. The region had no roads, only trails, and ranchers relied on moving their animals by river. The only route for cattle exports to Manaus and Amazonas was down the Rio Branco. Coudreau said that in 1885 forty boats a year went down the river, each carrying between fifteen and thirty-five animals – an annual total of about a thousand head. These were described as 'large heavy boats which have great difficulty passing the rapids of the Rio Branco, and then go on to Manaus in an even longer and harder journey'.

The round trip to Manaus took three months and it was very tough. The boats would return with a cargo of coffee, sugar, paraffin, tools, guns, cloth, and the many manufactured items unobtainable on the upper river – unless people traded with British Guiana. Conditions improved after the first decade of the twentieth century, when steam replaced human power.

During the rubber boom of the late nineteenth century, the rubber barons of Manaus badly wanted Roraima's meat. Continual calls for an end to the region's isolation resulted in the cutting, in 1893, of an 815-km trail (picada) from Manaus to Boa Vista; but this path was never used by cattlemen, and it rapidly reverted to forest. In 1900 a rough road was built from Boa Vista south across the plain to Caracaraí, by-passing the formidable Bem-Querer rapids that prevented large boats travelling along the Rio Branco. However, barge captains still preferred to risk the turbulent rapids to the labour of offloading their cattle onto the land portage.

When Hamilton Rice was in Roraima in 1927 he judged the area to be 'the best region of all the state of Amazonas . . . but "a paradise where people live poorly"' (Hamilton Rice 1928). Hamilton Rice wanted to build a railway from Manaus to Boa Vista. He offered to pay off the state's debt if he could have a sixty-year concession to exploit such a line and the land alongside it. Perhaps fortunately for the American explorer, the governor of Amazonas refused – since it is highly likely that Hamilton Rice would have ruined himself. Instead, in 1927 the Governor decided to reopen the old picada with a view to building a road along it. A consortium cut a 868-km path, using forty-five men working for twenty-one months (Oliveira 1929), but no progress was made with a road link. An official report in 1965 commented that Roraima still contained only 180 km of federal road from

Roraima's future

When Roraima became a state, it was felt that with its population of well over 100,000 and its economy in relatively good shape, Roraima was ready to take its place in Federal Brazil. Nevertheless, it is not certain that Roraima's economy can keep pace with the steady and inevitable rise in population and expectations of its people.

For a time in the late 1970s and early 1980s, the territory's economy was saved by timber exports. These were boom years for the Venezuelan construction industry, which discovered that the best wood for scaffolding and concrete formwork was a smooth white softwood called caferana (*Tachia guianensis*). This tree grows in abundance around the Bem-Querer rapids upstream from Caracaraí. Modern sawmills were developed to meet this demand, and the timber moved north along the new BR 174 link to Venezuela. Although caferana is a relatively cheap wood compared to tropical hardwoods, it was exported in large quantities. By 1980, timber exports had overtaken beef and cattle as Roraima's largest export. But this raw material is dependent upon market forces. With the collapse in the price of oil in 1986 and the Venezuelan economic crisis, demand for Roraima timber has decreased.

Roraima's cattle herd has declined from its recent peak in the late 1970s and early 1980s. Better methods of husbandry could sustain a further increase in cattle production but it is doubtful whether Roraima's weak pastures can feed many more animals. Scientific tests have shown conclusively that felling forests is not the answer. Soils under tropical forests are too impoverished to be transformed into good grassland (see Chapter 4).

There is potential for improved agriculture. An impressive increase in rice grown on seasonally flooded *várzeas*, as shown in Chapter 6, illustrates what can be achieved with good management and adequate financing. Here again, there is a limit to what can be produced and protection needs to be given to the endangered riverine forest.

The huge gold-rush of 1987 to 1990 brought thousands of *garimpeiros* flooding into the new state. They mined several tons of gold, but their strike was short-lived. The gold-rush brought prosperity to some service industries such as air taxis; but the arrival of so many new people put a strain on the region's fragile resources and little of the wealth that was generated found its way into the state's economy. The prospectors also severely threatened the Yanomami Indians and this raised the profile of the state within Brazil and internationally. The impact of *garimpagem* on Roraima's development is considered further in Chapter 7.

The gradual paving of the Manaus–Boa Vista highway will improve access to this remote part of Brazil. However, it will be many years before it can all be paved and the annual muddy blockages are at an end. This paving of Roraima's lifeline will improve the flow of imports and exports, but it will also attract more settlers. There will certainly be more industries to supply Roraima's needs, assuming a goal of self-sufficiency for the state, but these may create demand for more power than would have been delivered by the proposed (but never built) Paredão hydroelectric dam. The worry is that Roraima's growth will be at the expense of the magnificent rain forest reserves which protect, and are in turn protected by, its Indian population. If the present flow of immigrants turns into a flood, Roraima's delicately balanced economy and ecosystems could collapse.

Figure 2.4 Statue of *garimpeiro* in the main square, Boa Vista.
Source: Gordon MacMillan

3

MONITORING CHANGE IN LAND USE AND THE ENVIRONMENT

Tom Dargie and Peter Furley

INTRODUCTION

Reference was made in the previous chapter to the difficulties facing monitoring and policing of remote areas. The management of natural resources in regions such as Roraima is particularly conducive to the use of remote sensing (RS) techniques and geographical information systems (GIS). The lack of existing resource data, difficulty of ground access and isolation, combined with a dense forest cover and large area of protected reserves, provides remote sensing methods with both a challenge and an opportunity. Various techniques can be employed to collect data which would otherwise be impossible to acquire accurately or comprehensively.

A combination of RS and GIS approaches at different time periods, permits analysis of both spatial variation and temporal change. Despite the recent advance of ranching into forest areas and the intensification of land use in the grasslands and gallery forests in Roraima, there remain clear-cut patterns of natural savanna, forest and wetland, and an approximate coincidence of land use and vegetation zones. Furthermore, the pattern of flooding during the wet season, the control of topography over drainage, and the relationship of both to soil, is well suited to interpretation and analysis using a combination of RS and GIS methods.

Prior to the SLAR (Side Looking Airborne Radar) surveys in the early 1970s, there was little or no accurate information for most of the state. Projeto Radam generated volumes of data (Brasil 1975a, 1975b; 1978), mapped at the reconnaissance scale of 1:250,000, reduced for publication to 1:1 m (Almeida 1984; Furley 1986). Information under the headings of geology, geomorphology, soils and land evaluation, vegetation and land-use potential, provided the first comprehensive data available for natural resource assessment or development planning. Further assessment of the land potential for arable agriculture, ranching and forestry was provided in the Ministry of Agriculture's 1980 survey (see Chapter 1).

Subsequent to Projeto Radam, there has been relatively little further work

Table 3.1 Airborne radar and satellite flight characteristics of earth resource surveys used in the Amazon Basin

(a) SLAR (Side Looking Airborne Radar): 1973–early 1980s

System: GEMS 1,000,
  X-band frequencies operating at 9,375 MHz – 3.2 cm resolution 16 m
  swath width 37 km; covering 31,450 km^{2}h^{-1}
Main output:
  semi-controlled radar mosaics at 1:250,000
  infrared air photographs at 1:130,000
  multi-spectral air photographs (4 channels), selected areas at 1:70,000
  videotape: orthogonal to lines of flight at *c*.1:23,000; patchy cover of utilisable
    material
  altimetric data: profiles along lines of flight
  published volumes in series 'Natural Resources Survey', covering geology,
    vegetation, soil, geomorphology and land use potential.

(b) Satellite flight characteristics of Earth Resources Satellites (from UK National Remote Sensing Centre Guide to Earth observing satellites)

	NOAA AVHRR	LANDSAT MSS	LANDSAT TM	SPOT 1 & 2
Launch date	12-DEC-84(9) 17-SEP-86(10) SEP-88(11)	23-JUL-72(1) 22-JAN-75(2) 05-MAR-78(3) 16-JUL-82(4) 01-MAR-84(5)	16-JUL-82(4) 01-MAR-86(5)	22-FEB-86(1) 22-JAN-90(2)
Orbit	NPSS	NPSS	NPSS	NPSS
Altitude (km)	833–870	919	705	830
Inclination	98.7°–98.9°	99.09°	98.2°	97.7°
Coverage	82°N–82°S	82°N–82°S	81°N–82°S	81°N–81°S
Repeat cycle	12 hours	18 days	16 days	26 days
Equator crossing time	02.30(9,11) 14.30(9,11) 07.30(10) 19.30(10)	09.30(1,2,3) 09.45(4,5)	09.45	10.30
Spatial resolution	1.1 km[LAC] 40 km[GAC]	80 m	30 m 120 m[6]	20 m[XS] 10 m[PAN]
Swath width	3,000 km	185 km	185 km	60 km
Spectral resolution (wavelength in micrometers)	0.50– 0.68[1] 0.73– 1.10[2] 3.55– 3.93[3] 10.30–11.30[4] 11.30–12.50[5]	[ab] 0.50–0.60[1,4] 0.60–0.70[2,5] 0.70–0.80[3,6] 0.80–1.10[4,7]	0.45– 0.52[1] 0.52– 0.60[2] 0.63– 0.69[3] 0.76– 0.90[4] 1.55– 1.75[5] 10.40–12.50[6] 2.08– 2.35[7]	0.51–0.73[PAN] 0.50–0.59[1] 0.61–0.68[2] 0.79–0.89[3]

Notes: ( ) satellite number; NPSS: near polar sun-synchronous; LAC: local area coverage; GAC: global area coverage; [ ]: spectral band number; [a]: LANDSAT 1, 2, or 3; [b]: LANDSAT 4 or 5, band numbers renamed; [PAN]: SPOT panchromatic; [XS]: SPOT multispectral.

carried out in Roraima, either ground resource surveys at a more detailed scale or monitoring surveys utilising remote sensing. There have been agricultural and population censuses but these have not been related to the nature and pattern of natural resources. There has also been some work commissioned by FUNAI to monitor *garimpagem* and to undertake limited resource surveys in Indian areas. There has, however, been a regional analysis of the incidence of burning and deforestation from NOAA-AVHRR and LANDSAT satellite images. The federal agencies IBAMA and INPE have provided some interpretation of land-use change, particularly rates of deforestation, but inevitably this has been at a broad scale. The AVHRR surveys give a capability for frequent monitoring at a small scale – 1 km² (Cross 1990; Fearnside *et al.* 1990). The US earth resources satellites, LANDSAT MSS and TM and the French system SPOT have also been used either singly or comparatively (Nelson *et al.* 1987) (see Table 3.1). The Brazilian forestry agency (previously IBDF and currently part of the environmental agency IBAMA) has monitored deforestation at a more detailed scale, by manually overlaying images to give a time sequence, as illustrated earlier in Figure 1.13 (Brasil 1983).

Remote sensing and geographical information systems are particularly helpful in areas where ground information is sparse yet predictions have to be made rapidly for planning purposes; extrapolations need to be made from whatever data is available. As part of the Maracá Rain Forest Project, RS and GIS techniques were used to obtain data for areas too difficult to reach by ground survey and to provide a spatial database for interpretation and analysis. There were three principal objectives:

1 to provide a detailed vegetation map of Maracá Island from satellite images;
2 to provide a regional land cover assessment of the area around the Island, particularly encompassing the agricultural frontier between Maracá and Boa Vista to the east;
3 to compare the MSS data for 1978, representing one of the earliest clear images available, with TM data for 1985, being the latest suitable image available prior to fieldwork in 1987–88.

ANALYSIS OF PROJETO RADAM DATA AND CONSTRUCTION OF A GIS DATABASE

Projeto Radam was an extremely ambitious project, aimed at providing a reconnaissance level of data collection and interpretation. It generated rapid, broad-scale information in order to shape the path of development planning. Its tremendous achievement was to give the first comprehensive picture of natural resources in the Brazilian Amazon. Projeto Radam was inevitably generalised and demanded more detailed follow-up surveys which, as it turns out, could not be afforded and were never carried out except for

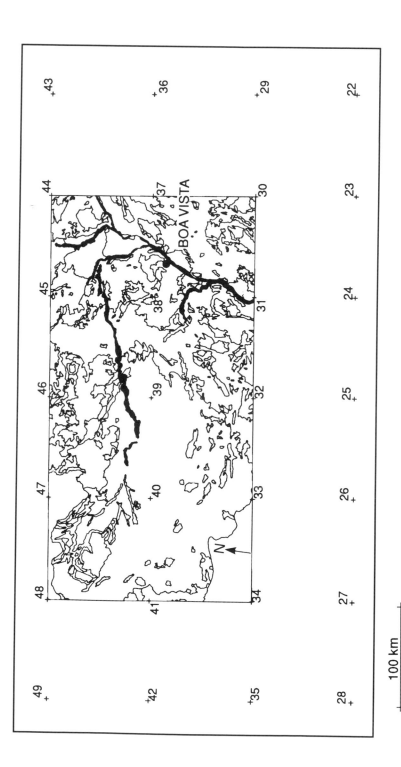

Figure 3.1 Extract from the GIS grid established for northern Roraima. Points represent a 100-km grid derived from the 1:100,000 topographic map series.

limited areas outside Roraima. As a result, the small scale of data available prior to the Maracá Project led to numerous inaccuracies at field level. For example, the soil work by Projeto Radam on Maracá Island was confined to one soil survey profile and one soil fertility site in an area with maximum axes of 70 km E–W by 30 km N–S (over 100,000 ha).

The survey does, however, provide a very useful baseline and context for resource evaluation. Consequently, a GIS for Roraima as a whole has been designed and 100 square km grids have been digitised into the database for analysis (Figure 3.1). Some of the errors inherent in this type of approach have been discussed in the context of northern Roraima by Downs (1991). Examples of the single attribute maps (or coverages) for Maracá Island are given in Furley *et al.* (1993). Examples of broader tracts of vegetation and soil distribution for the upper Rio Uraricoera (the curved path of which is clearly visible) are illustrated in Figures 3.2 and 3.3. A number of features stand out from this reconnaissance level survey, such as the approximately

RELIEF & DRAINAGE ATTRIBUTES IN THE TEPEQUÉM AREA,
DATABASE FOR RORAIMA

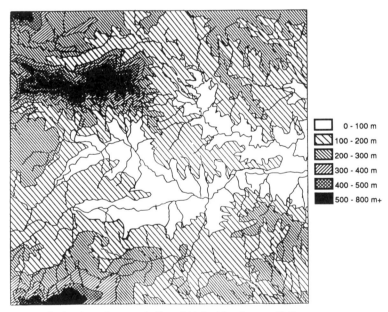

0 - 100 m
100 - 200 m
200 - 300 m
300 - 400 m
400 - 500 m
500 - 800 m+

Maracá Island can be seen in the mid left of the figure with the
Tepequém *tepuis* lying immediately to the north

Figure 3.4 GIS coverage of topography showing the Tepequém area. Maracá Island can be seen in the mid-left of the figure, with the residual hill mass (*tepuis*) of Tepequém lying immediately to the north.

Figure 3.5 Photograph showing the Tepequém *tepuis.*

N–S trend of the forest–savanna boundary (Plate 3.1), the recent incursions made by ranchers and smallholders into the forest, the density of the drainage pattern and the spectacular relief associated with the residual mountains or *tepuis*, notably the Tepequém area to the north of the island (Figures 3.4 and 3.5). The GIS analysis also permits overlay interpretation of one attribute upon another. Figure 3.6 illustrates the relationship of selected aspects of vegetation against soil type using the query facility of a relational database language.

The scale of Projeto Radam does not permit this type of analysis to be extended in any depth and the errors resulting from the limited data quality are considerable. The exercise was of value mostly to set up a working GIS which can be utilised and upgraded as more detailed and accurate data are made available in the future.

SOILS UNDERLYING DENSE TROPICAL FOREST,
GIS DATABASE FOR RORAIMA

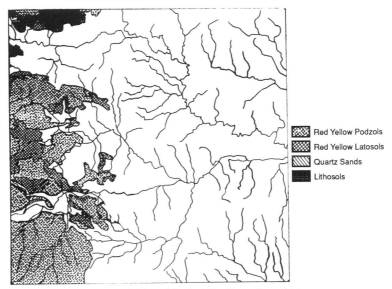

Red Yellow Podzols
Red Yellow Latosols
Quartz Sands
Lithosols

Maracá Island can be seen in the mid left of the figure (based on
digitization of the Projeto Radambrasil data)

Figure 3.6 Vegetation distribution against soil type using GIS overlay techniques.
Maracá Island can be seen to the mid-left of the figure. The overlay of vegetation,
in this case dense montane forest, on soils illustrates the technique of relating
variables.

ASSESSMENT OF LAND COVER AND VEGETATION
FROM LANDSAT MSS AND TM

Maracá Island

The principal objective of the RS analysis was the compilation of a map
showing the spatial distribution of vegetation on Maracá Island. There were
considerable problems of survey for an area of 101,000 ha, even with a large
team of scientists working in the field for over a year. Much of the area
could not be visited, although the overall density of observations and the
quality of the 'ground truthing' was very high for the Amazon region. The
detailed results for the botanical and other surveys have been published
elsewhere (e.g. Milliken and Ratter 1989, Milliken and Ratter forthcoming).
A black and white air photo mosaic (1:70,000) enabled a preliminary
geomorphological and drainage interpretation to be made (Eden and McGregor
1989), but it was evident that satellite imagery would be necessary to provide
ecological data for the larger part of the island.

Considering the area around Maracá Island and the nature of the Uraricoera river region, there are a number of striking features immediately apparent on the image (Plate 3.2). The first was the clear distinction between forest and savanna, the boundary of which ran approximately north–south and cut across the eastward side of Maracá. This represents the regional transition discussed earlier in Chapter 1. The change from forest to hyperseasonal (seasonally wet) grassland (*campo*) as well as to the more familiar dry *cerrado* is also evident (Furley and Ratter 1990; Thompson *et al.* 1992; Ross 1992; Ross *et al.* 1992). It is also apparent that there are other patterns representing smaller-scale distributions of vegetation.

Numerous problems became apparent in utilising the available LANDSAT MSS and TM images. Despite some 10 years of satellite passes, the number of images with little or no cloud cover was extremely limited. The MSS of 1978 and the TM of 1985 were amongst the best available and provided a 7-year period for comparison. For the purposes of producing a vegetation classification, the TM data set of 1985 was utilised as it was more up to date and the spectral and spatial properties of TM imagery are superior to MSS (Table 3.1). However, the 1985 data were not ideal, as the island itself was bisected by the orbit paths of the satellite; the eastern third being on PATH/ROW 232/58 and the western two-thirds being on PATH/ROW 233/58. This bisection caused problems of geometric accuracy between the images and, due to the time interval between the images (39 days), the spectral properties differed, primarily because of changing atmospheric conditions.

The geometric correction required in order to stitch the two sections together was limited by the quality of the ground control points which were taken from the largest-scale topographic maps available (1:100,000). A final correction was produced which gave a very favourable root mean square error of 27 metres for the western section and 29 metres for the eastern section. These figures are probably over-optimistic for the whole image, but the resulting fit of the two sections and the quality of registration with the available map data were very good under the circumstances.

Despite these difficulties, a vegetation zoning has been completed which can be used as a base map for future ecological research (Plate 3.3). The scale of the map can be adjusted according to the data input and a GIS database ensures that future information can be geo-referenced and thereby directly compared. The map output is currently produced by vector arcs using the ARC-INFO software package, which can be downloaded to a PC if funds become available for basic computer facilities at the research station in Boa Vista.

The spectral differences between the western and eastern sections were overcome by using a small section of the island which was in the overlap region of the two orbit paths. The assumption was made that, for the area of overlap for the period between the two images (39 days), no significant change had occurred in the spectral properties of the vegetation at canopy level. Therefore, any spectral difference was primarily due to atmospheric differences between the two dates. In order to correct the differences, a histogram-matching procedure was used, where the eastern section of overlap produced a reference histogram for each band to which the respective histogram of the western section was matched. The functions calculated from this procedure, which were more accurately known for the eastern sector, were then used to alter all the values for the western two-thirds of the island. A number of intermediate procedures, including filtering, ratioing and principal components analysis were then used in order to maximise the differentiation of the vegetation communities.

The classification procedure was essentially unsupervised, on account of the very limited ground truth available across the whole of the area. The unsupervised clustering procedure, generated 25 spectral clusters which were then used as spectral signatures in a minimum distance classification. The resulting classes were subsequently interpreted with the assistance of the field botanists. Interpretations were improved with the help of oblique aerial photography taken during the fieldwork on Maracá Island during 1987/88. However, the interpretation was limited by the timing of the fieldwork and the extent of the ground truth available.

For the final classification, the 25 classes were simplified to 13, two of these being associated with cloud cover. The classification attempts a comprehensive coverage of the island and there will inevitably be errors in extrapolations from the limited sites known in detail on the ground. The final distributions should be seen as a first attempt to produce a vegetation zoning for the island rather than providing the definitive map. Copies of the three colour prints of the vegetation classes can be obtained on request. A by-product of the image processing was the generation of a set of areal statistics for the various classes (see Table 3.2).

Table 3.2 Areas and proportions of vegetation classes for Maracá Island

Class	Area (ha)	Percentage
Semideciduous closed canopy forest	34,136.102	33.56
Intermediate forest types	14,709.961	14.46
Evergreen closed canopy forest	36,215.961	35.60
Open canopy forest types	2,979.090	2.93
Buritizal	1,855.260	1.82
Vazante vegetation types (a)	903.690	0.89
Vazante vegetation types (b)	414.360	0.41
Unflooded savanna	478.890	0.47
Shallow water/emergent vegetation	4,108.590	4.04
Trees overhanging water	1,899.810	1.87
Deep water	2,380.410	2.34
Cloud cover	1,020.420	1.00
Cloud shadow	621.630	0.61
Total	101,724.174	100.00

Determination of the spatial distribution and quantification of the vegetation classes provided the basis for a GIS analysis of the data. The map output can be generated in a variety of forms and at a number of different scales. As it is geo-referenced, the vegetation distribution can be compared with other data sets to determine relationships between vegetation and environmental variables such as soil or drainage. Furthermore, the GIS can be designed to include all the extremely diverse forms of biological collection and classification, although the lack of consistency in approach does lead to considerable problems in producing a readily understood and interactive spatial database (Maslen 1992). The analysis and map composition can be performed in a number of GIS software systems, for instance ARC/INFO or ERDAS, which can be mounted on IBM PC-compatible computers. Such a system could readily be installed at suitable locations at modest cost.

Land cover assessment

A second objective was to identify and classify land cover types over the broad area of approximately 200 km square from Boa Vista to Maracá. Plate 3.4 illustrates the classes identified on one-quarter scene LANDSAT TM in 1985. Clear patterns are evident and comparison of successive satellite passes could provide an effective monitoring capability of land-use change. This can also be quantified using a relational database and GIS.

With the availability of satellite data covering a significant period of inward migration and settlement in Roraima, an attempt was made to measure the physical growth of Boa Vista since the capital has attracted some 80 per cent of the population of the state (Turner 1989). The aim of Turner's study was, firstly, to calculate the growth of Boa Vista over the period 1978 to 1985 and, secondly, to assess the technical and methodological problems involved in this type of study. Using the same MSS and TM images which had been geometrically corrected and co-registered, it was found that the urban area increased by a figure of between 9.8 and 10.7 sq. km over the seven years. This represented an increase of 63 to 73 per cent or 9 to 10.5 per cent per annum, which is a very striking reflection of the demographic changes in the state.

There are several potential errors in this estimate. There was limited ground truthing available; however, the distinctive urban structure helped with location (Figure 3.7) and is even evident in Plate 3.4 despite the small scale. Further, the 1978 MSS imagery had a scale resolution that was unsatisfactory to act as a base line for comparison. In spite of these problems, with enhancement of the image it was possible to produce a reasonable estimate of the main urban area, although it probably missed most of the detail in the scattered semi-urban fringes and open shanty areas. Depending upon definition therefore, the analysis probably underestimated the already rapid rate of urban expansion. The MSS image was resampled

Figure 3.7 Aerial view of the centre of Boa Vista.

to 30 m pixel size (from 80 m) to provide direct comparison with the TM. Extracts of an area of 207 sq. km (571 by 403 pixels) covered the city and immediate surroundings. Not all of the city was included; for instance the airport and government owned land immediately to the south was excluded from the analysis. This was partly because the road separating the plots from the main urban area was exceptionally clear on both images. Finally, because the images were taken at different seasons, the riverine area was liable to vary. This was tested and, at least for the years examined, was not found to generate a significant difference (only 1 or 2 pixels).

In summary, both urban and rural change can be followed from the LANDSAT images, providing suitable cloud-free images can be obtained and that an acceptable level of matching between scenes of different dates can be provided.

Plate 3.1 Remote sensing image (LANDSAT TM) showing the regional forest–savanna boundary in north-west Roraima in the vicinity of Maracá Island. The lighter colours indicate the *cerrado* and cleared areas lying to the east of Maracá Island.

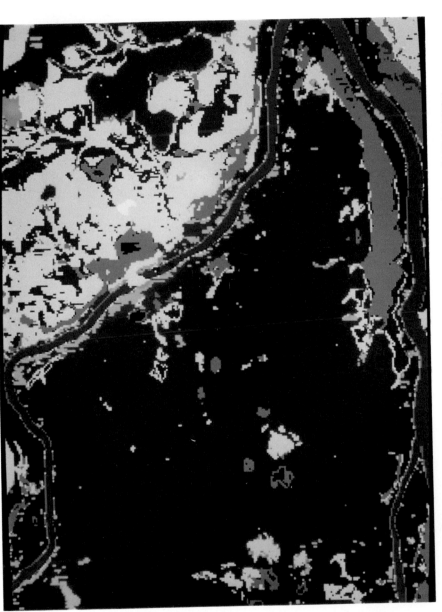

Plate 3.2 Vegetation of the eastern end of Maracá Island based on LANDSAT TM.

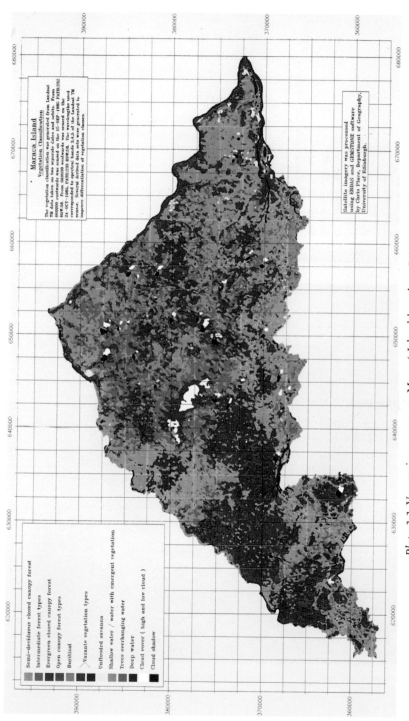

Legend:

Semi-deciduous closed canopy forest
Intermediate forest types
Evergreen closed canopy forest
Open canopy forest types
Buritizal
Vazante vegetation types
Unflooded savanna
Shallow water / water with emergent vegetation
Trees overhanging water
Deep water
Cloud cover (high and low cloud)
Cloud shadow

Maracá Island
Vegetation Classification

The vegetation classification was generated from Landsat TM data taken on two separate dates and orbits. From 060/060 overheads was scanned on the 15-SEP-1985 PATH:232 ROW:58. From 060/060 westwards was scanned on the 24-OCT-1985, PATH:233 ROW:58. The wavelengths used corresponded to spectral bands 3,4,5 of the Landsat TM sensor. Several derived data sets were generated to improve differentiation of vegetation classes.

Satellite imagery was processed using ERDAS and GEMSTONE software by Chris Place, Department of Geography, University of Edinburgh.

Plate 3.3 Vegetation zones on Maracá Island based on LANDSAT TM.

Water
Swamp
Wet Campo
Dry Campo
Cerrado
Forest
Forest
Forest
Cleared Forest
Bare Ground
Vazante
Cleared Campo & Cerrado
Cloud
Cloud Shadow
Unclassified

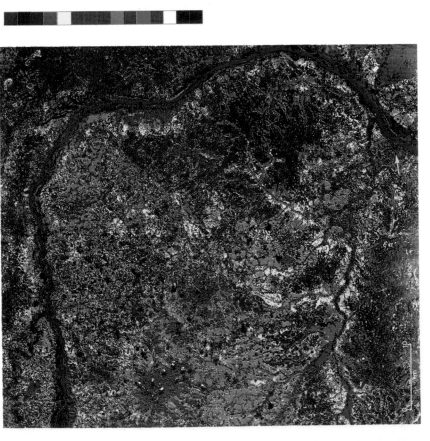

Plate 3.4 Land cover distribution for north-west Roraima from Boa Vista westwards to Maracá Island. The map is based on 1985 LANDSAT TM data and shows the complicated mosaic of nature and altered land use in the vicinity of Boa Vista – dominantly a *cerrado* area with patches of woodland, increasing in density westwards towards the forest edge.

MONITORING LAND COVER CHANGE

The capability of measuring land-use change using multi-date imagery is one of the principal assets of remote sensing. Of particular interest, in the region around Maracá, is the level of deforestation following the westward migration of the agricultural frontier and the related change in land use of both forest and savanna ecosystems (see Chapters 4 and 5).

Table 3.3 Image details and study area statistics

Image details			Study area	
LANDSAT 2 Multispectral scanner (MSS)			TM pixel total 11,608,000	
15 February 1978	249.58	Bands 4–7	Total area (based on 30 m pixel size)	
16 February 1978	250.58	Bands 4–7	10,447.2 km²	
Pixel size c.80 m				
			Western sector:	
LANDSAT 5 Thematic Mapper (TS)			TM pixel total	3,981,544
15 September 1985	232.58	Bands 3–5	Total area	3,583.4 km²
Pixel size c.30 m				
			Eastern sector:	
			TM pixel total	7,626,456
			Total area	6,863.8 km²

LANDSAT 2 (MSS) for 1978 and LANDSAT 5 (TM) data for 1985 were compared over a total area of nearly 10,500 km² (Table 3.3). The methodology included the creation of subscenes, replacement of missing scan lines in the 1978 data set, destriping the 1978 data to reduce the effect of 6-line banding caused by miscalibration of the sensor and, finally, geometric correction of the images. The establishment of training areas involved training on cover types from ground data and from inference, and gives an idea of the probability of accurate classification (see Table 3.4).

RESULTS AND DISCUSSION

The land use survey has been summarised in Plate 3.4, with 13 land cover units identified plus cloud, cloud shadow and unclassified units. Transition matrices and cover totals were calculated and a number of observations can be made:

1 Open cover types, which include water, swamp, wet and dry *campo* (grassland), bare ground and *cerrado* (shrubby savanna), make up almost 60 per cent of the study area. They reveal a complex pattern of inter-type change between the two image dates. This pattern probably represents a seasonal flux between types and doesn't affect the overall balance for cover classes.
2 There is a very strongly marked seasonal change. Much of the dry season *cerrado* changes to wet season swamp and wet *campo*, and confirms

81

Table 3.4 Image correction and training methods, classification and image comparison

---

Geometrical correction
1 Image-to-map (UTM) 1985 TM data fitted to co-ordinates from IBGE 1:100,000 map sheets; cubic convolution, 30-m pixel size; accuracy +/− 50 m.
2 Image-to-image 1978 MSS images converted to corrected 1985 TM image with resampled 30-m pixels:
 (a) western 1978 sector +/− 4 pixels accuracy;
 (b) eastern 1978 sector +/− 1 pixel accuracy.
Training methods
1 Cover types trained on ground data.
2 Cover types trained on inference:
 cloud, cloud shadow;
 burnt *campo* and *cerrado*;
 cleared *campo* and *cerrado*;
 cleared forest;
 forest types (single-band density slicing);
 1978 swamp, wet *campo* and dry *campo* mosaic.
Training areas per cover type
1 Several, scattered throughout the image.
2 Minimum total of 1,000 pixels to calculate mean variance and inter-band co-variance.
Classification: Maximum likelihood – 95 per cent probability of pixel belonging to one cover-type class, accepting that there will be unclassified pixels (unlike the Island vegetation map). At least two cycles of training area definition, purification and classification were needed to produce acceptable classified images.
Image comparison: Pixel counts of changes (transition patterns) were derived by comparing the class of each 1978 pixel with its 1985 equivalent, and then by assembling results into transition matrices for eastern and western sectors.

---

the field observations that much of the *cerrado* area is seasonally water-logged or very wet.
3 There are large unexpected reversals within the pattern of seasonal fluxes in land cover. These reversals suggest that improved training is still required to clarify seasonal change, probably by redefining the transition boundaries for swamp, and wet and dry *campo* areas.
4 Forest losses are significant despite the relative remoteness of the area. The raw statistics are given in Table 3.5.

Table 3.5 shows that there has been a significant loss even in the remote western margins of the forest edge (over 4 per cent) and a very large loss has occurred in the more developed eastern sector (nearly 30 per cent) between 1978 and 1985. This gives an overall 10 per cent loss from the total 1978 baseline forest cover. Losses in the west appear to be mainly in upland and valley forests and seem to be mostly associated with colonisation, usually with a nucleated settlement core and gridiron feeder road system joined linearly to other parts of the region. This is well demonstrated by Alto Alegre and other sites studied during the land development programme

Table 3.5 Loss of forest between 1978 and 1985 (raw statistics)

	Western sector	*Eastern sector*	*Total area*	*Percentage of total area*
Forest (area km²)				
1978 MSS	1848.6	525.5	2374.1	22.72
1985 TM	1773.5	375.7	2149.2	20.57
change	−75.1	−149.8	−224.9	
% change	−4.06	−28.51	−9.47	
Cleared forest (area km²)				
1978 MSS	137.2	46.6	183.8	1.76
1985 TM	160.0	189.6	349.6	3.34
change	+22.8	+143.0	+165.8	
% change	+16.62	+306.87	+90.21	

(Figure 3.8; see also Chapter 5). In the east, the very high rates of forest depletion relate to valley and flood-plain locations and this is most likely to be the valuable riverine or *várzea* forest (see colour plate section). The forest losses appear to be associated with the growth of commercial agriculture such as rice or horticultural developments (see Chapter 6).

These figures are immensely significant given the perceived isolation of the area. However, they also need to be treated with caution, because the calculations for forest loss and cleared land do not always balance (especially in the western sector), and some of the losses can be attributed to misinterpretation of the original forest area (as wetland swamp, *vazantes* or non-forested storm flow lines, or wet savanna). A further possible source of error is that the 1978 MSS data may have overestimated the forest area, because of the difficulties in interpreting certain forest features such as open lines (*veredas*) of buriti palm (*Mauritia flexuosa*) and storm lines (*vazantes*) at the 80 m pixel scale. This also applies to the eastern sector, where much of the cleared forest – as defined in 1985 – may actually be swampy areas of *várzea* converted to rice production.

Interestingly, large areas of the cleared forest identified in 1978, are recorded as forest in 1985. This may indicate abandonment of cleared areas and would be consistent with the high levels of pasture abandonment noted over eastern Amazonia (Serrão and Toledo 1988). Furthermore, and only evident in the 1985 image because of its recent nature, there is considerable conversion of *cerrado* (whether wet or dry *campo*, arboreal, shrub or open *cerrado* interspersed with patches of bare ground), into dry rice and improved pasture. This represented only 0.35 per cent of the area examined but, from field observations, appeared to be expanding rapidly.

CONCLUSION

The use of remote sensing has permitted the production of an island vegetation map which, although the result of prediction for most of the

Figure 3.8 A comparison of seven years' development at Alto Alegre, 1978–85.

western sector where the least field botany was conducted, provides nevertheless an immensely valuable baseline for future biological research. Remote sensing has generated the first detailed land cover assessment for the north-western region of Roraima. Further, comparison of land cover and urban environmental changes over time has given a graphic view of the pace of disturbance in an area often perceived as relatively isolated from characteristic Amazonian development. The establishment of a geographical information system for Maracá Island, also permits manipulation of the current database and provides the framework for the more comprehensive and detailed resource surveys which will be necessary. The quality and availability of such data is crucial for the planning of development in the state, even if confined to eliminating the more obvious mistakes elsewhere.

Looking to the future, the next generation of sensors is likely to provide a greater range of detail at larger scales. The use of synthetic aperture radar, as used in the European Space Agency Programme (ERS-1) or the American TREE (Tropical Rain Forest Ecology Experiment) programme, will provide data on forest morphology, plant biochemistry and be able to trace the incidence and spread of disease. The TREE experiment, for example, located in the rain forest areas of southern Mexico, Guatemala and Belize, collects imaging spectrometer and imaging radar data from NASA aircraft, using Jet

Propulsion Laboratory sensors with synchronous ground surveys. Whereas LANDSAT TM gathers data on 7 spectral bands, AVIRIS (Airborne Visible and Infrared Imaging Spectrophotometer) gathers images electronically in 220 separate bands. Since the concentrations of plant component materials (such as cellulose, lignin and pigments) differ between species, it is likely that subtle differences in vegetation may be detected by such sensors – well beyond the current resolution and interpretation of LANDSAT used in the present study. The imaging radar (AIRSAR) also permits measurement of the total amount of vegetation (biomass) and water tied up in the vegetation. There is little doubt that the tools available for biological assessment and development planning will be greatly increased over the next few decades and that accurate description and monitoring will become widely accessible within the areas they are needed.

With the problems of distance, inaccessibility and limited technical support, and faced by rapid rural and urban change, remote sensing provides a cost-effective means of providing data on the types of land development described elsewhere in this book.

ACKNOWLEDGEMENTS

Grant assistance was received from the Ford Foundation/Royal Geographical Society, the Carnegie Trust for the Universities of Scotland, the Baring Foundation and from the Gilchrist Trust. Grateful acknowledgement is also made to Christopher Place, University of Edinburgh, for assistance with image processing, and to the logistical support given to the Royal Geographical Society's Maracá Rain Forest Project, and by SEMA (now IBAMA) and INPA in Brazil, and especially to Sr Guttenberg Moreno.

4

DEFORESTATION AND THE ENVIRONMENT

Michael Eden and Duncan McGregor

INTRODUCTION

The precise extent of cleared forest land in Amazonia is not at present known, but probably totals some 8–12 per cent of the historic forest cover (Neto 1989). As far as the state of Roraima is concerned, the current extent of clearance is relatively low, but increasing rapidly as has been shown in the previous chapter. It is thus of relevance to examine the present and likely future impact of forest clearance on the environment of Roraima. The impact will be discussed at two levels, namely, the local and the regional. Local effects will be considered mainly in relation to forest clearance for cattle ranching and for small-scale agriculture. Attention will focus, firstly, on changes in soil chemical fertility on cleared forest land and, secondly, on soil physical changes and associated slope dynamics. Specific data will be presented from an area of forest clearance near Maracá Island in northern Roraima. The regional effects of forest clearance, which are currently rather speculative in character, will be considered in the light of prevailing environmental conditions and of relevant findings elsewhere.

In view of the relatively recent and limited character of forest clearance in Roraima, there is at present little evidence which illustrates environmental impacts. The general effects, however, are likely to be similar to those reported from parts of Amazonia where more substantial clearance has occurred. These include areas of cattle ranching in eastern Pará, northern Mato Grosso and parts of the Andean *oriente* (Falesi 1976; Fearnside 1980; Hecht 1981; Toledo and Serrão 1982; Nelson *et al.* 1987, Uhl *et al.* 1988) and of small-scale agricultural colonisation along the Transamazon highway in central Pará, the Vilhena–Pôrto Velho highway in Rondônia and other similar roads (Moran 1981; World Bank 1981; Smith 1982; Coy 1987; Woodwell *et al.* 1987). In respect of the regional impact of forest clearance, attention has primarily focussed on possible climatic and hydrologic feedbacks (Fearnside 1985a; Salati *et al.* 1986; Myers 1988) and on loss of biodiversity (Sioli 1980; Myers 1986; Taylor 1988).

As elsewhere in Amazonia, the picture emerging in Roraima is that forest

clearance generally results in relatively favourable, short-term land productivity, but that, unless agrosystems are devised that are adaptive in
ecological as well as economic terms, it is very difficult to sustain productivity.
Widespread land degradation can result, and impoverished secondary plant
growth replaces the original tropical moist forest. On the broader scale, it
is likely that large-scale deforestation in Roraima will contribute to the
positive feedbacks, physical and biological, that are anticipated for Amazonia
as a whole. An effective forest conservation strategy for the area is thus
required.

FOREST VEGETATION

Roraima is covered by a range of vegetation types, among which tropical
moist forest, savanna, steppe savanna and pioneer formations are the most
widely encountered (Brasil 1975a, 1975b; 1978). Of these, tropical moist
forest is the most extensive, occupying an estimated 15.6 million ha or 68
per cent of the state (see Figure 1.12), although the forest cover is variable
in character, primarily in response to regional variations in climate, particularly
rainfall, and relief.

Since the present study is concerned with the environmental impact of
contemporary deforestation in northern Roraima, a relevant starting point
is the ecological status of the Amazonian forest itself. In this respect, it has
often been previously assumed that the forest was an ancient and stable
ecosystem, which as a function of its accumulated species diversity, was
relatively resistant to disturbance. Latterly, however, it has been widely
accepted that substantial changes have occurred in Amazonian climatic
conditions during the Pleistocene period and that these have induced major
changes in forest distribution (Haffer 1969; Prance 1985; Whitmore and
Prance 1987). In parallel, the forest system has come to be seen as dynamic
and fragile in character and vulnerable to disturbance (May 1975). The
substantial loss of Amazonian forest that has occurred in recent decades,
mainly through pioneer colonisation by migrant farmers and cattle ranchers,
may be cited as evidence of such vulnerability.

While there is no doubt that substantial anthropic deforestation has
occurred in recent decades (Fearnside 1986b), it should be pointed out that
the Amazonian forests, including those of Roraima, have long been occupied
and exploited by an indigenous population, without their having suffered
any apparent substantial or lasting damage. In recent millennia, the main
indigenous impact has been in the form of shifting cultivation, which
is still practised in many parts of the region, including large tracts of
Roraima. The limited impact of this form of cultivation has commonly
been related to the adaptive status of the agrosystem itself. In this respect,
emphasis has been placed on the diverse compositional and complex
structural characteristics of the crop community which has been seen

advantageously to simulate that of the natural forest (Harris 1971; Meggers 1971). Latterly, attention has also focused on the status of indigenous shifting cultivation as a successional or integrated 'field and fallow' system, whose temporary and localised clearance for cropping is readily followed by natural woody regeneration (Uhl *et al.* 1982; Eden 1987; Eden and Andrade 1987). As long as the integrity of the 'field and fallow' system is maintained, shifting cultivation remains adaptive and compatible with the inherent fragility of the forest ecosystem.

When forest disturbance occurs on a larger scale, however, as is often the case in areas of contemporary cattle ranching or peasant colonisation, where blocks of thousands of hectares may be cleared, the regenerative capacity of the forest is impaired (Gómez-Pompa *et al.* 1972, 1991). This would be less worrying if the cleared land were destined for permanent exploitation, but, in many cases, the chances of this happening are slight and the land in question is often abandoned after a few years. Weedy shrubs and woody pioneer species readily re-establish themselves, but restoration of a primary forest is less assured. This reflects the reproductive characteristics of the primary forest trees, whose seeds tend to be large, rather short-lived and dependent on animal dispersal (Gómez-Pompa *et al.* 1972; Kubitzki 1985). Such species can colonise small gaps, either natural or associated with shifting cultivation, but their re-establishment across extensive tracts of cleared forest land, where seed sources are lacking and forest animals scarce, is greatly impeded. Gómez-Pompa *et al.* (1972) suggest that primary forest species 'are incapable of recolonizing large areas opened to intensive and extensive agriculture'. This perhaps overstates the case in the light of the apparent ability of the Amazonian forest to recover from periodic refugial contractions during the Pleistocene period (Haffer 1969), but it rightly emphasises the difficulty, on a human timescale at least, of achieving large-scale regeneration of primary forest.

While there is genuine cause for concern over maladaptive, contemporary exploitation of Amazonian forest, the situation in Roraima is relatively favourable in the sense that the scale of forest clearance is currently less than in many parts of the region. According to Tardin *et al.* (1980), the area of cleared forest in Roraima in 1975 was only 5,500 ha. By 1978, the area had more than doubled to 14,375 ha, but still represented a mere 0.2 per cent of the total estimated clearance in Brazilian Amazonia at the time. In the last decade, clearance in Roraima has substantially increased, with very serious effects in places (see p. 83), but is still low by Amazonian standards. Even so, migration to the state is rapidly increasing and popula-tion numbers have risen significantly in recent times, particularly in areas beyond the immediate savanna hinterland of the capital, Boa Vista. In parallel with this, pioneer colonisation and forest clearance are extending. As elsewhere in Amazonia, this generally results in favourable land pro-ductivity for a few years, but in the longer term such exploitation is unlikely to be sustainable or avoid causing substantial, positive environmental feedbacks.

LOCAL EFFECTS OF FOREST CLEARANCE

The effects of forest clearance can be viewed at two separate levels – local and regional. These effects are partly related to the mode of clearance and to the subsequent use made of the land. The main categories of use are small-scale agricultural colonisation, cattle ranching, extractive logging and mineral exploitation. In the past, deforestation resulting from mineral exploitation has been minimal in Roraima, but recent *garimpeiro* activity in Yanomami areas of the upper Uraricoera basin has had damaging, local effects on riparian and other forests, as well as on the Indians (Margolis 1988). Similarly, despite bouts of greater felling (see Chapter 2), extractive logging has not had a major impact in Roraima, with the state recording the lowest levels of log production in Brazilian Amazonia (IBGE 1988). Despite the optimism expressed by IBGE (1981) of an enlarged market for timber in Venezuela, the remoteness of the region from large markets would seem at present to preclude logging as a major economic option.

Small-scale agricultural colonisation has taken place, notably at Taiano, Alto Alegre and north-west of Caracaraí. Detailed discussion of the nature and extent of agricultural colonisation may be found elsewhere in the volume. However, agricultural colonisation *per se* represents a relatively small percentage of permanent land use compared with the area under cattle ranching. For example, IBGE (1981) reports the area purely under agriculture in the municipalities of Boa Vista and Caracaraí as about 267,000 ha, compared with about 1.5 million ha in pasture (1975 census figures).

Most of the area of pasture in Roraima is concentrated on the savanna plains around Boa Vista, where cattle were first introduced in the late eighteenth century (see Chapter 2). Only in recent decades have there been any significant attempts to extend ranching onto adjacent forest land, notably along the banks of major rivers such as the Uraricoera and the Mucajaí. Apart from data presented in this volume, little detailed evidence exists for the rate of forest clearance in Roraima, although a broad picture at a small scale is collated by the national space agency (INPE) and the monitoring section of IBAMA. Equally there is little information on the effects of clearance on the chemical and physical properties of forest soils, but parallels may be drawn with similar areas elsewhere in the Amazon.

Several environmental problems arise at the local scale when forest land is cleared for pasture. It has at times been assumed that such pastures, when well-managed, will maintain themselves in good condition for many years (Falesi 1976) and this view has in the past led to official encouragement of large-scale pasture development (Fearnside 1980). The shortcomings of such development have latterly been elaborated (Serrão *et al.* 1979; Fearnside

1979, 1980; Hecht 1981, 1985; Buschbacher 1986; Buschbacher *et al.* 1986; Eden, 1990; Eden *et al.* 1991), but the process continues in many areas.

In many cases, pasture productivity is favourable for an initial 4 to 5 years, but thereafter pastures frequently deteriorate. The broad pattern is of soil nutrient depletion, deterioration of soil physical conditions and acceleration of slope erosional processes. According to Serrão *et al.* (1979), the primary constraint on soil productivity in eastern Amazonia is the availability of soil phosphorus. Phosphorus levels are relatively high immediately after initial forest clearance and burning, but availability of the nutrient declines to minimal levels over time. With respect to soil physical conditions, deterioration is apparent in the increasing bulk density of pasture topsoils, which suffer reduced infiltration rates and increased sheetwash erosion (Hecht 1981).

Weed invasion is also a serious problem in areas of derived pasture. Woody weeds rapidly invade pasture land, especially when the introduced forage grasses lose initial vigour and time-consuming and expensive manual weed control becomes necessary (Dantas and Rodrigues 1980; Hecht 1981). Burning of weedy pastures is often practised but, although disposing of existing weed material, it does little to arrest general decline in pasture quality. The status of the pasture is also aggravated by overgrazing, which helps accelerate soil deterioration and weed invasion. In spite of its damaging impact, overgrazing is often an attractive economic option since cleared, albeit degraded, land can readily be sold at a profit and the proceeds re-invested in further clearance of forest land (Fearnside 1980; Hecht 1981).

Soil chemical conditions

The soils of Roraima exhibit significant variability at local and regional levels. As yet, only a reconnaissance survey has been undertaken over most of the state and, particularly in more remote areas, no more than preliminary soil data are available. Numerous soil types have been identified and have been summarised in Chapter 1. The soils are generally variable in texture, commonly as a function of parent materials, but are mostly of low chemical fertility.

In general, Amazonian soils, including those of Roraima, are relatively acid, dominated by low activity clays and contain minerals of low solubility in subsoil layers (Furley 1990). The majority of available nutrients are concentrated in the vegetation itself and in the uppermost 10 to 15 cm of soil. Organic matter in the soil is particularly important because of its ability to provide exchangeable cations for uptake by roots. The supply of nitrogen from natural sources is also largely dependent on soil organic matter while, particularly in more dystrophic soils, released nutrients such as phosphorus appear to be directly re-cycled through mycorrhizal systems (Jordan 1985).

When the forest is cleared for either agriculture or pasture, leaching and

Figure 4.1 Derived pasture of *Brachiaria humidicola* at Fazenda Patchuli. The site was cleared some 19 months previously, cropped for maize and rice, and then converted to pasture.

erosion lead to loss of nutrients. Clearing the land by burning also involves significant losses of organic material by combustion, returning some nutrients to the topsoil but also causing loss of nitrogen and sulphur to the atmosphere.

Forest clearance and the establishment of pasture follow a relatively uniform pattern in northern Roraima. Forest trees are generally cut by axe or chainsaw during the dry season from September to March and, after partially drying out, the resultant debris is fired. Initially, the cleared land is devoted to subsistence crops, which are normally planted to coincide with the first rains in April or May. The main crops are maize and dry rice, with subsidiary crops like beans, manioc and banana. Grass cuttings are also planted among the food crops. In earlier times, *colonião (Panicum maximum)* was the most common pasture grass in use, together with *jaraguá (Hyparrhenia rufa)*, but since about 1980 *quicuio de Amazônia (Brachiaria humidicola)* has been widely adopted (Figure 4.1).

Initial pasture quality is generally favourable in the area, with woody weeds effectively controlled by manual weeding and regular burning. In subsequent years, pasture productivity declines and older pastures are in a very degraded condition. Such pastures commonly support a low cover of native grasses and variable densities of herbaceous and shrubby weeds. The sequence of nutrient changes is illustrated by data from Fazenda Patchuli, located in an area of current clearance for pasture to the south of Maracá Island (Figure 4.2). The sites sampled were chosen to represent a time sequence from forest, through a cleared cultivation stage to progressively older pastures up to 12 years and into adjacent natural savanna. The pastures, including savanna, are generally grazed on a rotational basis. On the derived pastures, stocking levels are usually 1–2 animals per hectare. Soil samples were collected from sites over a distance of some 2 km along an access track that extended from undisturbed forest, through cleared forest land into natural savanna. Relatively uniform conditions of parent material and of undulating convexo-concave topography characterised the area. Topsoil samples (0–10 cm) were collected at 11 sites, 2 in the forest, 7 in forest land cleared for pasture, and 2 in savanna. At each site, 4 replicate samples were obtained. These were individually collected at randomly selected points within an area of approximately 30 × 30 m in the central part of each site. All samples were obtained at the end of the rainy season (August–September 1987), some 6 months or so after the annual cutting and burning period in the area.

Laboratory analyses were undertaken as follows: organic carbon was determined by the Walkley–Black method; exchangeable Ca, Mg, Na, H and Al were extracted with 1 M KCl and exchangeable K with 0.02 M HCl, and were measured by atomic absorption spectrophotometer or flame photometer; available P was determined using the Bray 1 method; pH was measured in a 1:2.5 suspension in H_2O and dilute $CaCl_2$ (Black et al. 1965).

At Fazenda Patchuli, soil parent materials are relatively uniform in texture with sandy loam to sandy clay loam topsoils overlying sandy clay to clay at depth. The soils are provisionally classed as Ultisols. The soil chemical data (Table 4.1) show some initial effects of conversion of forest to pasture. This is apparent in spite of the fact that sampling of first-year sites occurred some 6 to 8 months after cutting and burning of the forest, by which time the land had already been cropped for maize and rice, and much of the original soil nutrient boost from burning already depleted. Even so, a substantial increase in pH is evident at first-year sites although there is subsequent decline. It is noticeable, however, that even older pasture sites retain a higher pH than adjacent forest sites. Notable increases in exchangeable calcium and magnesium and a slight increase in potassium also occur, but again there is subsequent decline. Initial increase in available phosphorus also occurs, but, under prevailing soil acidity, available phosphorus levels as a whole are still very low and below that regarded as adequate for most crops (Landon 1984). As Serrão et al. (1979), Fearnside (1980) and others indicate, phosphorus seems to be the most limiting factor in Amazonian pastures, resulting in poor growth irrespective of other fertility indicators. This is presumably also the case in the present area where, in spite of initial

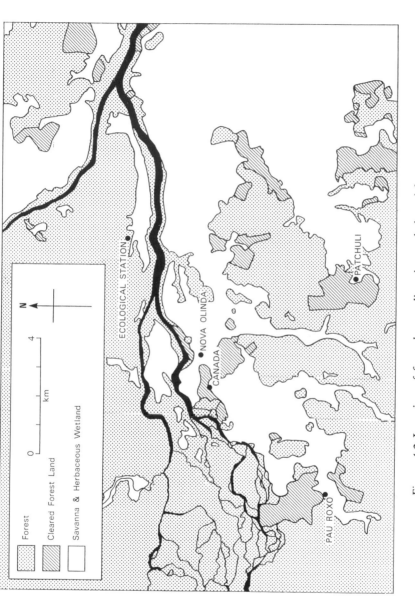

Figure 4.2 Location of *fazenda* sampling sites and cleared forest areas.

Table 4.1 Analytical results for topsoil samples (0–10 cm) at Fazenda Patchuli, Roraima

	Number of samples	Organic carbon (%)	pH (H₂O)	pH (CaCl₂)	Exchangeable (cmol kg⁻¹)						Available P (mg 100 g⁻¹)
					Ca	Mg	K	Na	Al	H	
Forest	8	1.58	4.6	4.0	0.81	0.86	0.11	trace	0.36	0.97	0.36
Cleared Forest											
year 1 cultivation	8	1.22	6.4	5.7	3.78	2.28	0.17	0.01	0.30	0.20	0.61
year 2/3 pasture	8	1.93	5.6	4.9	2.16	0.70	0.22	0	0.06	0.29	0.50
year 4/6 pasture	8	2.14	5.5	4.7	1.79	0.47	0.20	trace	0.10	0.36	0.41
year 12 pasture	4	1.36	5.4	4.6	1.08	0.33	0.09	0	0.08	0.29	0.35
Savanna	8	1.09	4.9	4.1	0.22	0.08	0.07	trace	0.48	0.36	0.31

increases in available phosphorus, minimal levels are generally re-established within a few years of clearance and burning.

In general, the present data substantially confirm the findings reported elsewhere in derived Amazonian pastures, both in respect of soil chemical conditions and of weed invasion (Falesi 1976; Serrão *et al.* 1979; Fearnside 1980; Hecht 1981; Moran 1981; Cochrane and Sanchez 1982). As indicated, the main problem appears to be low available phosphorus levels which, in the absence of fertiliser inputs, are likely to limit pasture productivity and carrying capacity. There is some evidence elsewhere in the vicinity of Maracá Island that reduced stocking levels generally improve the condition of pastures (Eden *et al.* 1991), but explicit pursuit of this strategy would obviously reduce short-term profitability. Equally, it is possible that fertiliser applications would improve pasture quality, but, in isolated areas like northern Roraima, such inputs are costly and unlikely to be economic in the foreseeable future. According to Serrão and Toledo (1988), the introduction of improved grass and legume forages with low fertility requirements would be an appropriate method to lower nutrient needs and hence management costs. Suitable forage varieties are not yet widely available, but may become so as the region is developed. For the present, the best strategy for ranchers in northern Roraima is to exploit the nutrient capital of the standing forest; regrettably therefore, every incentive exists to continue clearance and maintain the supply of newly established pastures.

Short-term soil physical changes

Empirical data on soil physical changes resulting from deforestation are generally lacking, but, as with soil chemical conditions, it is unlikely that patterns of change will be markedly different from those elsewhere in Amazonia. In respect of short-term physical changes, removal of the protective vegetation cover immediately exposes the soil to raindrop impact. Since high intensity storms with large raindrop size are characteristic of the region, rainfall is potentially highly erosive (Brandt 1988) and also has a high potential for soil compaction (Martins *et al.* 1991). In addition, soil physical changes are affected by local soil textural and structural conditions.

Topsoil bulk density is a soil variable that integrates the climatic and edaphic factors mentioned above. Bulk density data are available for areas of forest conversion to pasture in several parts of Amazonia. The data indicate that bulk densities increase relatively rapidly after forest clearance (Hecht 1981; Nicholaides *et al.* 1984; Dias and Nortcliff 1985; Martins *et al.* 1991). This is mainly due to direct soil compaction, but may also reflect loss of topsoil organic matter. Soil compaction in turn affects infiltration rates and is a critical influence on soil erosion.

Data from similar environments elsewhere in Amazonia suggest that rates of soil erosion are low under forest, and increase with clearance (Smith 1976;

McGregor 1980; Fearnside 1986b; Forsberg *et al.* 1989). Erosion rates vary with the type of clearance, being in general higher on agricultural land and lower under pasture. This is essentially a response to the degree of protection afforded by the plant cover, and has been confirmed experimentally. For example, in the middle Caquetá basin of Colombia, McGregor (1980) showed that rates of erosion are higher in fields cultivated by traditional methods than in derived pasture with a continuous grass cover. However, topsoil compaction under pasture leads to increased overland flow, raising the potential for gullying where wash is concentrated, as, for example, by road cuttings or by drainage/irrigation works. It is expected therefore, that the physical response of the forest soils in Roraima will differ somewhat between agricultural and pastoral land use.

In order to examine the physical response of forest soils to clearance in the Maracá area, data were obtained on soil texture and topsoil bulk density. Three core samples (115 cm³) were taken in topsoil (0–10 cm) at each sample site, and bulk density determined after oven-drying to constant weight at 105°C. Particle size analysis was undertaken on topsoil and subsoil (75–100 cm) samples using the pipette method (Black *et al.* 1965).

Data on rates of soil erosion under forest and following clearance are available for nine 15 m by 6 m experimental plots on Maracá Island (Ross *et al.* 1990), and qualitative observations were made, by the present authors, of soil erosion phenomena in the vicinity of Fazenda Patchuli.

The main physical change observed in the Patchuli area is in topsoil bulk density (Figure 4.3). Relatively low values under forest (1.0 to 1.2 g cm³) increase in the early years after clearance. At first, the rate of increase is relatively rapid, rising to 1.25 to 1.3 g cm³ in first year clearance (sampled 6 months after clearance) and to 1.4 g cm³ in the second year field. Bulk density values of about 1.5 to 1.6 g cm³ are common in older pasture sites,

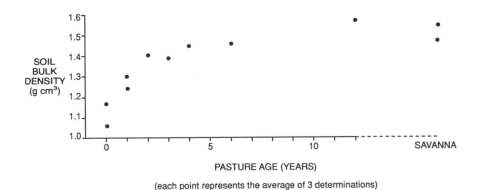

Figure 4.3 Bulk density of topsoil samples (0–10 cm) at Fazenda Patchuli, Roraima.
Source: After Eden *et al.* 1991.

96

Table 4.2 Topsoil textures, Fazenda Patchuli

	Number	Sand (%)	Silt (%)	Clay (%)
Forest	8	72	8	20
Pasture				
1st year	6	71	7	21
Young	10	67	9	24
Old	8	70	9	22
Savanna	8	74	7	19

approximating to those recorded in adjacent savanna sites. Similar trends of bulk density increase are apparent at other sites in the vicinity of Maracá Island (Eden *et al.* 1991).

The increase in bulk density has occurred in the absence of mechanisation, clearance being by traditional 'slash and burn' alone. Also, increases in bulk density are not related to soil texture (Table 4.2), indicating that initially, before pasture is established, compaction largely takes place by raindrop impact, and that compaction by trampling becomes more significant when pasture is established.

Soil erodibility, an indicator of erosion potential, is affected by soil organic content. Soils with less than 2 per cent organic carbon (equivalent to about 3 per cent organic content) are generally regarded as erodible (Morgan 1986). At Fazenda Patchuli, organic carbon initially builds up in the topsoils sampled (Table 4.1), and then declines, in older pastures, to low levels similar to those of adjacent savanna. However, the Patchuli soils contain, at best, just above 2 per cent organic carbon, and may therefore be regarded as potentially erodible, especially in older pastures.

Forest clearance on Maracá Island has been shown, as elsewhere in Amazonia, to increase erosion. Data collected by Nortcliff *et al.* (1989) indicate that substantially increased run-off and soil losses occurred in experimental plots totally cleared of forest vegetation. Over a 3-month period, a fortyfold increase was measured in mid-slope soil loss between 3 plots with undisturbed forest and a further 3 adjacent cleared plots. Whilst this result undoubtedly reflects steeper slope angles and the presence of a lateritic horizon at shallow depths in the mid-slope sites, the potential effects of forest clearance on erosion rates in the Maracá area are clearly demonstrated.

At Fazenda Patchuli, the presence of colluvium mantling lower slopes on pasture sites and over adjacent valley floors strongly suggests that the conversion from forest to pasture is associated, particularly in the initial phase, with increased rates of sheetwash erosion. Splash erosion, which will contribute to sheetwash, was also observed to have taken place during rainfall events, grass and timber debris being coated with fines within a few centimetres of the soil surface, particularly following short, heavy downpours.

In the wider Maracá area, field observation indicates that many areas of savanna are mantled by sandy sheetwash (Figure 4.4), whereas sheetwash is

Figure 4.4 Gullied sheetwash east of Fazenda Patchuli.

less frequently encountered under forest. This may, however, reflect relatively limited field observations under forest, as there are no clear textural differences between the forest and savanna soils sampled (Figure 4.5).

Long-term soil physical contrasts

The short-term physical consequences of forest clearance in the Maracá area are increased soil compaction and increased rates of erosion. In view of the fact that clearance and associated land degradation have been confined to recent decades, the long-term outcome of forest clearance in the area cannot be determined directly. However, significant geomorphological contrasts exist between forest and savanna in the Maracá area. The savanna is a more degraded environment, which may thus constitute an analogue for the long-term outcome of forest conversion to permanent pasture.

In order to examine this hypothesis, it is necessary to examine the nature of the geomorphological materials, processes and forms under both forest and savanna in the study area, which determine the likely long-term consequences of deforestation. They will also be useful to establish the likelihood of a savanna landscape resulting, with time, from forest clearance. This necessitates consideration of much longer timescales than hitherto in this account.

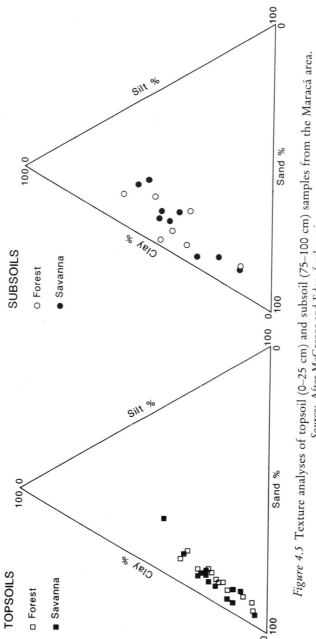

Figure 4.5 Texture analyses of topsoil (0–25 cm) and subsoil (75–100 cm) samples from the Maracá area.
Source: After McGregor and Eden, forthcoming.

Figure 4.6 The flat savanna summit near Nova Olinda.

A model may be set up to characterise savanna and forest terrains, and to bring out the broad contrasts in geomorphological forms and materials that exist between them. Areas of Amazonia in which forest appears to have remained throughout the Quaternary may be expected to exhibit a low degree of plinthite hardening (Eden *et al.* 1982). Incipient concretions are found within soil profiles, possibly within a water table fluctuation zone at shallow depth, but indurated layers are infrequently encountered. Soil parent materials comprise more or less weathered bedrock. Summits are usually rounded rather than flat, while slope forms are generally convexo-concave. Sheetwash processes occur, but are less efficient than in the savanna. Geomorphologically, a relatively undegraded environment is indicated.

In contrast, areas of Amazonia in which savanna appears to have prevailed intermittently throughout the Quaternary commonly exhibit stripped profiles, with hardened *in situ* plinthite at or close to the surface in many places. Flat summits, often with associated gravel deposits and flanking sheetwash accumulations are also common, giving many parts of the savanna a rather angular appearance (Figure 4.6). Gravels generally consist of a mix of angular to subangular quartz and ironstone and rounded to sub-rounded pisoliths (Figure 4.7), and are presumed to have been derived, at least in part, from the break up of hardened plinthite (duricrust). These phenomena

Figure 4.7 Lateritic gravel on a summit north of Maracá Island.

appear to confirm extreme geomorphological degradation, although it must be noted that the savanna landscape varies widely, and that materials other than gravels or colluvium may predominate locally.

The differences outlined above largely reflect the differing physical processes of forest and savanna, and describe the situation observed in the Maracá area. Significant differences in slope form and degree of plinthite hardening (duri-crust formation) are repeatedly encountered in transects of as little as 100 m across the present-day forest–savanna boundary, and bear little relation to variations in parent material. In this part of Roraima therefore, open vegetation seems to have been sufficiently stable for significant geomorphological contrasts to develop (McGregor and Eden, forthcoming). Within the present-day forest on the other hand, there appear to be tracts which were previously more open with inclusions of *cerrado* (Furley and Ratter 1990).

The present savanna landscape in the Maracá area is thus characterised by stripped profiles and exposure of hardened plinthite, the outcome of geomorphological processes acting in the absence of a forest cover. Against this background, the expected long-term effects of forest clearance appear to be analogous. Nicholaides *et al.* (1984) indicate that laterisation hazard arises 'only when the subsoil is exposed'. In their view 'the key is to prevent the soft plinthite in the subsoil being exposed by erosion of the topsoil. It is only then that irreversible hardening takes place.'

With respect to forest clearance, as indicated by the data of Nortcliff *et al.* (1989), a topsoil 'stripping' process would almost inevitably follow exposure of the soil to accelerated colluvial action. Plinthite, as has been encountered under forest in the Maracá area, would be exposed at the soil surface and would either be eroded or would harden on exposure. Hardening would be enhanced locally by lower precipitation and by lowered water table levels, such as might follow regional reductions in rainfall totals following extensive deforestation. Depending on the relative rates of erosion and hardening, a duricrust could eventually be formed where plinthite is present; and with continuing denudation the duricrust could locally form higher ground with colluvial infilling at the slope foot. Slope forms could then change from convexo-concave to more angular, with the resultant landscape closely resembling existing savanna, both as regards slope form and materials.

As yet, the above process is little advanced in the forested area of Maracá. However, on recently cleared land at Fazenda Patchuli, it is noticeable that forest clearance has led to accelerated sheetwash processes and the accumulation of colluvial materials in the valley bottoms. These conditions are ideal for the exposure and hardening of plinthite formed close to the top of the soil profile. The critical question is how much plinthite is present within the existing forest zone; and whether sufficient exists for hardening to become a potential long-term problem following forest clearance.

Hardened plinthite is generally associated with the more seasonal areas of Amazonia and has been encountered by the writers at numerous locations northwards from Boa Vista (Figure 4.8) and in savannas of the Maracá area. The occurrence of 'concretionary and latcritic soils' has been noted in these areas and elsewhere (IBGE 1981).

In the forest around Maracá, plinthite consists of iron-rich clay and incipient iron concretions associated with few, generally fine, quartz gravels. Only reconnaissance investigations of the soils have been undertaken so far, but frequent occurrences of plinthite have been reported. Nortcliff and Robison (1988), for example, encountered plinthite at several sites during surveys on the Maracá Island and noted that there is evidence of a plinthite plateau across the south-east of the island. It is not possible to estimate the extent of plinthite under forest, although significant occurrence is indicated.

The current tendency of scientists working in the Amazon Basin is to reduce the emphasis given to plinthite with respect to land development. Certainly, the more pessimistic views regarding the dangers of its conversion to duricrust (McNeil 1964; Goodland and Irwin 1975) are no longer acceptable. As Cochrane and Sanchez (1982) indicate, the areal extent of soils with subsoil plinthite in Amazonia is limited to only about 4 per cent. However, this is still equal to about 21 million ha, and, where plinthite does occur, it may be a significant factor in constraining land use.

The contemporary forest boundary is the focus of land-use change in

Figure 4.8 Concretionary pisolithic gravel (plinthite) exposure, north of Boa Vista.

Roraima, as demonstrated in the next chapter. At present the principal change is conversion of forest to pasture, but with some conversion to agriculture. Attention must therefore be drawn to the contrasting surface materials and slope forms observed in the forest zone and in adjacent savanna areas. Following the argument presented above, it may be true that areas presently under forest, and appearing to offer rather better conditions for pasture, would eventually degrade to the status of the savanna area, and potentially in the longer term lead to the production of duricrust.

REGIONAL EFFECTS OF FOREST CLEARANCE

Large-scale forest clearance in Amazonia is likely to involve regional environmental feedbacks. Concern over physical feedbacks has focused on climatic and hydrological as well as geomorphic variables, while biological feedbacks have mainly been discussed in terms of loss of biodiversity. Since the scale of clearance in Amazonia, and especially Roraima, is still relatively limited in overall terms, there is as yet little, if any, direct evidence of the regional effects of deforestation. Nor is it clear what scale of clearance is required to induce such effects or, by implication, what level of conservation is necessary to avoid or minimise them. The intention here is to review possible regional feedback effects, assessing their implications for Amazonia in general and Roraima in particular.

Climatic and hydrological effects of forest clearance

It is widely assumed that large-scale clearance of forest will affect regional climatic and hydrological conditions and induce changes that are significant in human terms. Earlier suggestions that widespread deforestation would upset the global oxygen balance have been discounted but the impact of deforestation on the heat balance and hydrological cycle continues to give cause for concern. The physical processes involved are complex and predictions of change very difficult, but it is anticipated that regional and even global environmental effects will occur.

Many investigations in Amazonia have concentrated on the relationship between deforestation and rainfall (Molion 1976; Marques *et al.* 1977; Lettau *et al.* 1979; Salati *et al.* 1979, 1986). According to Molion (1976) and Salati *et al.* (1979), the supply of water vapour in Amazonia is initially derived by air flow from the Atlantic Ocean, but approximately 50 per cent of regional rainfall is attributed to local evapotranspiration and re-precipitation. The presence of a forest cover, which reduces rates of terrestrial run-off, is assumed to facilitate this local recycling and thus increases rainfall levels.

Conversely, if deforestation occurs, terrestrial run-off is likely to increase and evapotranspiration and rainfall be correspondingly reduced. According to Sioli (1980), replacement of a freely transpiring forest by a 'steppe-like' cover might lower evapotranspiration to one-third of the current level, causing a significant reduction in rainfall total and increased rainfall seasonality. At what scale of clearance such effects would become apparent is less clear. Fränzle (1979) considers that localised forest clearance 'limited to the areas adjoining transport routes and settlements' would have little influence on precipitation levels. However, more than localised clearance will occur if current rates of Amazonian land colonisation continue, and significant hydrological feedback is probable.

Consideration has also been given to the effect of forest clearance on atmospheric carbon dioxide. The latter is currently increasing as a result of net carbon release from fossil fuels as well as from decomposing and burning forest biomass and soil organic matter. Such release is generally expected to induce changes in both global and regional climates. During the last 100 years, atmospheric CO_2 has increased by 20 to 25 per cent, giving a present concentration of some 350 ppm. The net annual atmospheric increment commensurate with the current rate of increase is 2.9×10^9 tonnes C y^{-1} and continuing additions at this rate will probably double the atmospheric concentration at some time during the next century (Hare 1980; Walsh *et al.* 1981; Detwiler and Hall 1988).

How significant Amazonian deforestation is to this process is uncertain.

The amount of carbon stored in the biomass of Amazonia is estimated at 115×10^9 tonnes (Sioli 1980, 1985), but the net flux from this pool will depend as much on the nature of the replacement land cover as on the extent of forest clearance itself. The situation is further complicated because the dynamics of the global carbon cycle are imperfectly understood, making it difficult to determine what proportions of released forest carbon pass to the atmospheric and the oceanic pools (Richey et al. 1980). Nevertheless, global forest clearance probably now contributes 20 to 30 per cent of the atmospheric CO_2 build-up, with Amazonian clearance probably responsible for at least a third of that total (Dobson et al. 1989). Resultant modifications to the atmospheric 'greenhouse effect' are expected to cause rises in global temperature and sea level, as well as changes in rainfall.

The overall impact on Amazonian climate is uncertain. Modified surface albedo and evapotranspiration will occur as a result of deforestation, while local as well as general increments of CO_2 will continue to pass to the global atmosphere. Forest clearance and burning are also causing a net flux to the atmosphere of trace gases like nitrous oxide (N_2O) and possibly methane (CH_4), and are raising aerosol levels, all of which, directly or indirectly impinge on the heat balance and may affect climate (Fearnside 1985b; Joyce 1985; Seiler and Conrad 1987; Pearman and Fraser 1988). While the precise prospect is unclear, drier and possibly warmer conditions may develop within Amazonia on a timescale of decades.

As previously indicated, forest clearance is likely to cause accelerated surface run-off, especially on steeper slopes, and will in turn lead to increased fluvial discharge and sediment load. Even if total rainfall declines, such feedbacks are likely to persist. During drier periods of the year, river levels may not be greatly modified, but peak flood levels will certainly increase. Flood responses of this kind are reported elsewhere in the tropics (Myers 1988) although, in respect of Amazonia, relevant empirical evidence has not yet been produced (Nordin and Meade 1982; Richey et al. 1980). Such feedbacks are of particular concern in Roraima because of the pronounced relief of its northern headwater zones. The impact of modified fluvial regimes will be most evident in the várzeas. Increased peak flooding will extend the area of seasonal inundation, putting settlements and cultivated land at risk. River sediment levels will increase and patterns of channel erosion and deposition vary.

Biological implications of forest clearance

Widespread clearance of primary forest in Amazonia will also lead to massive extinction of plants and animals and consequent loss of genetic resources. Any biome is prone to damage when disturbed, but tropical forest is particularly vulnerable because of its composition and dynamics. Not only is it exceptionally diverse and thus prone to multiple extinctions, but also many of its species are specialised in character and intolerant of habitat disturbance. Evolutionary factors, including the creation of Pleistocene forest refugia (Haffer 1969), have resulted in a localised distribution of many organisms, which are correspondingly vulnerable to local forest clearance (Prance 1977; Myers 1980). No precise estimate is possible of the likely scale of plant and animal extinctions in Amazonia, but predictions suggest that, if large-scale clearance continues, some 0.5 million species may disappear from all tropical moist forests by the end of the century (Myers 1977; Furley 1993). In Roraima, the scale of extinction will depend on the extent and location of clearance, but it can be assumed that corresponding losses will take place.

Losses of Amazonian genetic resources are undesirable for many reasons. Firstly, it will limit the capacity of the forest to reproduce itself. Pioneer plant species capable of initiating regeneration on cleared forest land will generally survive, but primary forest species, which are biologically more vulnerable, may not be available to sustain the return to maturity. The vulnerability of these species relates to the lower availability of their seeds, reflecting basic germination and dispersal characteristics as well as the sparse distribution of individuals of each species (Gómez-Pompa *et al.* 1972, 1991).

Secondly, the extinction of Amazonian species represents a loss of resources of direct value to humans. Numerous plant and animal products have already been identified and commercially exploited in the forest, but many other species of potential value exist there. The latter are vulnerable because their utility is often unrecognised or only locally known. They include many potential cultigens of value for the production of food, fibres and fodder (Tosi and Voertman 1964). The number of Amazonian forest trees, for example, that produce edible fruit or nuts is immense and, although a few like cacao (*Theobroma cacao*) and Brazil nut (*Bertholletia excelsa*) are already of major commercial importance, many others have unrealised potential as food products (Benchimol 1988).

Also of potential value are phytochemical products, or extractives, which exist in individual plants or taxa and have widespread industrial and other uses (Goldstein 1979). Such products include latex, oils, waxes, resins, dyes, tannins and other natural compounds. Latex, an emulsion of hydrocarbons and water is familiar as the raw material for the production of rubber from *Hevea brasiliensis*, but many other species of comparable potential exist in the forest and warrant investigation.

Forest plants also contain chemical compounds of medicinal value. Lovejoy and de Padua (1980) point out that the Amazon basin is one of the best places to seek such compounds, because the evolutionary processes that induce their development have been aided by the constant natural competition in the forest between plants, insects and other organisms. Natural products of this kind have been widely used by indigenous populations and some have been exploited by modern society. The traditional remedy ipecac

(*Cephaelis ipecacuanha*), which contains alkaloid substances, has been used in the treatment of amoebic dysentery, bronchitis and bilharzia (Myers 1984). Curare, derived from a range of alkaloid-bearing plants and used as a hunting poison, contains an active ingredient that has long been employed as a muscle relaxant in surgery (Mors and Rizzini 1966; Reis Altschul 1977). *Alexa canaracunensis*, a forest legume occurring in Roraima, is a source of the alkaloid castanospermine which, in laboratory experiments, exhibits activity against the AIDS virus (Lewis and Owen 1989). Meanwhile, the screening of other forest plants needs to continue in the search for natural compounds that will serve either as the direct constituents of medicines or drugs, or provide a starting point for their laboratory synthesis.

Thirdly, a loss of Amazonian species would put established forest products at risk. Forest species that have been brought into cultivation are still dependent on the genetic resources of their wild relatives and primitive cultivars for breeding purposes. Even when major plant improvements have been made in respect of energy and nutrient conversion or of stress resistance, continuing plant breeding is required and depends on the availability of a large and diverse gene pool. This applies as much to established commercial species, like rubber and cacao, as it does to any new domesticates, plant or animal, that may emerge from the forest. In the case of staple food crops, much less attention has hitherto been paid to plant breeding, but any future improvements will depend on the availability of traditional cultivars. As contemporary land colonisation extends, this material is as vulnerable to genetic erosion as are wild forest species, and needs to be conserved.

CONSERVATION PLANNING

While damaging environmental feedbacks are expected to result from extensive deforestation, their precise scale and impact are not easily predicted. This is particularly so in respect of the atmospheric system whose likely response to large-scale clearance is difficult to anticipate (Hare 1980; Sioli 1980). The biological effects of clearance are more predictable and, on such grounds alone, the case for substantial forest conservation can readily be made.

In this respect, some consideration has been given to the required extent and distribution of national parks in Amazonia. On the basis of island biogeography theory, Myers (1979) has suggested that conserving 1 per cent of the Amazonian moist forest might safeguard 25 per cent of the region's species. If 10 per cent of the forest area were conserved, 50 per cent of its species might be saved, while conserving 20 per cent of the forest area should safeguard most of its species. These data are no more than 'an informed first guess' (Myers 1979), and will in any case depend on how the reserves in question are distributed. In respect of Roraima, 20 per cent of the forest area represents some 3.1 million ha. Substantial reserved areas of forest and other vegetation have previously been proposed for conservation (see Table 4.3), notably by Projeto Radambrasil which nominated 12.3 million ha, or 53 per cent of the territory, for this purpose (Brasil 1975a, 1975b, 1978). Currently, however, there is only one official reserve in Roraima, namely, the Ecological Station of Maracá which occupies <1 per cent of the forest area (SEMA 1977). A further area for savanna protection has also been designated to the south and west of Caracaraí. Following extended negotiations, the Yanomami Indian National Park has recently been legally demarcated (Holden 1979; Margolis 1988).

Even if around 20 per cent of the forest area of Roraima is effectively established as reserved land, it cannot be assumed that the remaining forest should be available for clearance, since far more than a 20 per cent forest cover is needed for adequate protection of climatic and hydrological systems. However, not all the extra reserved land need be under natural forest, since forest reserves containing tree-based exploitation systems like agroforestry or mixed-forest silviculture can, if properly managed, provide effective physical protection. The total area of reserved land needed to achieve general physical and biological protection is not known. In the past, it has been suggested that 50 per cent of any environment should be retained as 'natural environment' (Odum and Odum 1972), but current concern over increasing

Table 4.3 Proposed and existing reserved land in Roraima

(a) Cited by Pandolfo (1974):	
Parima Forest Reserve (designated 1961)	1,758,000 ha
Rio Branco National Forest (proposed)	2,000,000 ha
Total	3,758,000 ha
(b) Proposed by Projeto Radambrasil (Brasil 1975, 1978):	
Roraima National Forest	2,401,600 ha
Rio Negro National Park	4,845,600 ha
Serra Parima National Park	2,506,000 ha
Serra Pacaraima National Park	1,073,000 ha
Lago Caracarana National Park	40,300 ha
Pedra Pintada National Park	8,400 ha
Ecological Stations (including Maracá)	1,409,600 ha
Total	12,284,500 ha
(c) The Atlas de Roraima (IBGE 1981) reiterated the reserved areas proposed by Projeto Radambrasil and identified as Indian reserves:	
Yanomami (Catrimani basin)	est. 1,500,000 ha
Makuxi reserve (Surumu basin)	est. 600,000 ha
(d) Reserves currently identified:	
Ecological Station of Maracá (designated 1976)	92,000 ha
Yanomami Indian National Park (designated 1991)	9,400,000 ha
Total	9,492,000 ha

atmospheric CO_2 and other 'greenhouse gases' and their likely warming effects on the global climate may soon be translated into demands for even higher levels of forest conservation (Nisbet 1988; Eden 1990; Prance 1990).

In spite of the uncertainty, designating national parks and forest reserves in Roraima and elsewhere is arguably easier than maintaining them. Many examples exist in Amazonia of the invasion of reserves by road builders, ranchers, peasant farmers, miners and mining companies (Anon. 1984, 1986; Fearnside and Ferreira 1984). In Roraima itself, the process is perfectly illustrated by the recent invasion of the Yanomami Reserve by *garimpeiros* (Margolis 1988; Schwarz and Rocha 1989; see also Chapter 7). Considerable effort is clearly required to protect such areas. Some direct policing is necessary, but effective, long-term protection will only be achieved if there is a parallel broad transition from pioneer land colonisation to more adaptive and sustainable modes of land use (Eden 1990).

CONCLUSION

At present, Roraima is experiencing rather less forest clearance than many parts of Amazonia, although its impact is very localised. However, the population and the scale of disturbance are now increasing more rapidly than in the past. Recent environmental impact is most evident in respect of *garimpeiro* activity in the north-west of the region, but local forest clearance for cattle ranching and small-scale agriculture is occurring in scattered locations elsewhere. Hitherto, few data have been acquired on the specific environmental impact of such activities, but as presently indicated, their local effects are often damaging in character and the activities themselves not easily sustained in the longer term. This is certainly the case with local ranching activities in the vicinity of Maracá Island. Initial pasture quality is generally favourable, but quickly declines. Nutrient levels are low, and in particular low levels of available phosphorus limit pasture productivity. Weed invasion further degrades the status of older pastures, and pastures are sustainable only at very low stocking densities. Soil physical degradation is apparent as pastures age, particularly in terms of increased soil compaction and increasing rates of sheetwash erosion, though it is not clear to what extent this affects pasture growth. In the longer term, it appears likely that significant physical degradation would result from permanent establishment of pasture, with the resultant rangeland resembling savanna in its physical characteristics.

At a regional level, the predicted positive environmental feedbacks of large-scale forest clearance have also been discussed. Both physical and biological feedbacks are involved and, although their likely effects are at this stage difficult to gauge, a strong case exists for developing a comprehensive conservation strategy for Roraima. Preliminary work in this direction has already been undertaken by Projeto Radambrasil (Brasil 1975a, 1975b, 1978) and the time is now ripe for effective implementation of their general approach.

ACKNOWLEDGEMENTS

Grant assistance was received from the Ford Foundation/Royal Geographical Society (MJE/DFMM), the Carnegie Trust for the Universities of Scotland (DFMM), and the Central Research Fund of the University of London (MJE). Grateful acknowledgement is made of logistical and other assistance provided by the Royal Geographical Society's Maracá Rain Forest Project, and by Brazilian Government Agencies, notably the Secretaria Especial do Meio Ambiente (SEMA), the Empresa Brasileira de Pesquisa Agropecuária (EMBRAPA). Particular thanks are due to Nelson Vieira for assistance in the field, to Margaret Onwu for laboratory assistance, and to Ron Halfhide and Justin Jacyno for cartographic work. Plant identifications were made at the Royal Botanic Gardens, Kew and Edinburgh.

5

FOREST CLEARANCE AND AGRICULTURAL STRATEGIES IN NORTHERN RORAIMA

Luc Mougeot and Philippe Léna

INTRODUCTION

So far, the discussion has considered historical development and rates of forest clearance (Chapters 2 and 3), and has assessed detailed relationships between land-use change and soil (Chapter 4). We now turn to specific colonisation settlements where the problems of forest disturbance are seen in the light of individual and household agricultural strategies, particularly those of the smallholder.

Rural settlement in Roraima can be attributed to a series of related general trends:

1 Increasing spatial mobility of people within the state of Roraima and influxes of land-seekers from other regions of Brazil. This includes immigrants from existing Amazonian settlement areas such as Ji-Paraná, Jarú, Ouro Preto and Ariquêmes in Rondônia or Gleba Jauaperi on the Perimetral Norte (Maciel da Silveira and Gatti 1988: 57).
2 Publicity in the late 1970s about state programmes of land distribution and grain production in Roraima, supplying growing local urban markets. This coincided with expansion and improvement in local road communications and the demonstrative effect of sponsored and private agricultural colonies and livestock estates in southern Roraima (Jatene 1984: 27–8; Maciel da Silveira and Gatti 1988: 49–53).
3 Modernisation of traditional livestock activities, largely through the planting of pastures in cleared forest, as ranchers perceive the need to improve the quality of their livestock. There appears to be an increased tendency for preferential clearing of denser forested patches over intensification of existing pastures, a feature which is now well established in southern Amazonia (Fearnside 1986b).

Environmental constraints are at the heart of a whole range of producer strategies which enhance family security by adapting systems of production. Forest clearance is part of such strategy and will be examined from the viewpoint of the viability of production units. This should afford an

111

understanding of the farm-level rationale for converting forest into other land uses, as well as advantages and inconveniencies of the resulting deforestation in localities where farms were surveyed.

STUDY LOCATIONS

Three localities were selected for survey: Vila Brasil, Taiano and Tepequém (Figure 5.1). These are representative of different natural and socio-economic environments in the region and characteristic of the zone of forest–savanna transition.

Vila Brasil is located in a zone of extensive savanna in contact with the forest, where there is a mosaic of grassland (*campo*), low shrub (*campo sujo*) and arboreal and shrub savanna (*cerrado*), gallery forest (*mata ciliar*), islands of forest and high uninterrupted forest cover, which can be dense evergreen or semideciduous. Since the first waves of colonisation, the traditional form of land development here has been extensive cattle ranching. It now has been practised for more than a century and shows some signs of modernisation in recent years. Sixteen ranchers were interviewed on their *fazendas*.

Lying to the south-east of Vila Brasil, Taiano was one of the first agricultural colonies (1953) established by the territorial, now state government, to supply the capital (see Chapter 2). This was achieved at Taiano by farming an island of forest endowed with more fertile soils than those typical of the region (Brasil 1975a and Chapter 1). Today small-scale commercial agriculture is hindered by a lack of land available for expansion. The extensive clearings where government-subsidised grain crops used to be grown have now been engulfed by a dense secondary regrowth. The removal of this regrowth has proved to be very labour-demanding and producers have not managed to establish a sustainable beef cattle industry. The area consistently suffers from seasonal water deficiency. Many of the plots have been subdivided and producers lack the means to accumulate or attract the capital that they require to switch to cattle-raising. Twelve farmers were interviewed on their smallholdings, in hilly terrain with semideciduous natural forest and with shrubby re-growth on lower-lying meadows.

The colony of Tepequém lies to the north-west of Vila Brasil and is mostly carved from a dense patch of high forest at the forefront of agricultural colonisation thrusting westwards. It has repopulated an area which was previously an outpost of placer-mining. Twenty-two farmers were interviewed on *fazendas* in the foothills and on the plateau of the Serra do Tepequém.

Altogether, a total of 50 farmers were interviewed and a survey was made of their respective plots. These surveys took place in the dry mid-season (period of clearing, felling, food scarcity). Half of the farmers were then re-interviewed during the following rainy season (period of rice harvesting, with damaging impact of rains on local road communications). This second

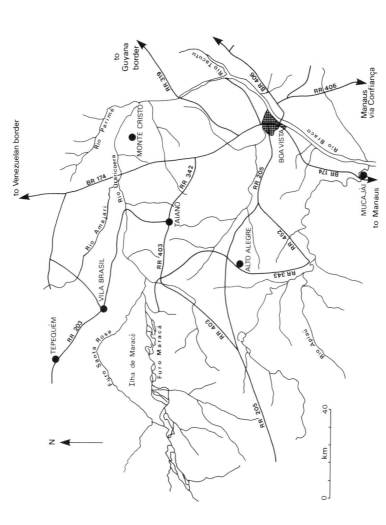

Figure 5.1 Location of survey sites. Land colonisation projects to the north and west of Boa Vista: Monte Cristo, Alto Alegre, Taiano, Vila Brasil and Tepequém – the last three being the focus of study in this chapter.

survey was made to check the accuracy and update the initial survey. It permitted a seasonal comparison to be made. An assessment was then possible of the constraints and discrepancies between expectations and achievements in late 1987.

Data were collected under the following headings: (a) the migrational and occupational history of the informants; (b) household composition and kinship; (c) past and current land use on the informants' holdings; (d) monthly pattern of rural activities (agriculture, livestock raising, hunting, fishing, forest product extractivism); (e) perceived environmental constraints (diseases, labour shortages, difficult transport conditions, price rises, food scarcity, need to borrow, etc.) during the past agricultural year; (f) organisation of work and use of manpower; (g) social practices (affiliation to syndicates, churches, co-operatives); (h) patterns of consumption.

FOREST CLEARANCE AND CONSTRAINTS ON AGRICULTURAL USE

In several previous chapters it has been shown how forests have been felled at a growing rate over the last 25 years in Roraima. This trend was confirmed during the detailed surveys of the three study areas selected, but with interesting variations.

On the basis of the field interviews it seems that some 14,000 out of the 31,000 ha of land which the 50 producers claim to own were under forest cover originally. Up until the early part of 1987, some 25 per cent of this cover (3,507 ha) had been cleared and converted into other uses, either by the informants themselves or by previous occupants of the farm areas. Interviewees alone were responsible for 55 per cent of this clearance (1,947 ha). Of these 1,947 ha, 15 per cent (or nearly 300 ha) had been felled during the agricultural year of 1986–87 (Table 5.1).

If the average clearance rate reported since early plot occupancy was to remain constant, plots taken as a whole in the three areas would be totally deforested within around 64 years (at Taiano this period would be reduced to 6 years). However, this forecast is based on historical averages which do not fully account for more recent increases in the pace of clearance. If one divides the area still under forest cover by the amount of land cleared in the year 1986–87, the number of years required for total clearance to occur falls to under 40 years. The estimated time to total clearance varies at different localities – from 43 at Tepequém to around 4 years at Taiano. It should be noted that the mean area felled in 1986–87 was itself already larger than the mean area of land cleared annually by the informants since arrival.

These estimates assume, of course, that the average size of annual clearings will remain constant beyond 1987 (Table 5.2), and nothing is certain about this assumption. Our own observations suggest a continued increase in the extent of forest clearing, for at least some years to come. There are many

Table 5.1 Estimates of forest clearance in the forest–savanna transition of northern Roraima: 1986–87

	Vila Brasil	Taiano	Tepequém	Total
Informants	16	12	22	50
Plot area claimed by informants (ha)	21,872	668	9,141	31,681
Declared plot area under forest cover originally (ha)	6,603	654	7,973	15,230
As a % of claimed plot area	30.2	97.9	87.2	48.1
Total declared forest cover area cleared by informants and previous occupants (ha)*	2,067†	545	895	3,507
As a % of declared plot area under forest cover originally	31.3	83.3	11.2	23.0
1 Cleared by previous occupants only (ha)	763	345	452	1,559
– As a % of declared total area under forest cover originally	11.6	52.8	5.7	10.2
– As a % of total declared forest cover cleared	36.9	63.3	50.5	44.5
2 Cleared by informants only, until Feb. 1987 (ha)	1,304	200	443	1,947
– As a % of declared plot area under forest cover originally	19.7	30.6	5.6	12.8
– As a % of total declared forest cover	63.1	36.7	49.5	55.5
3 Cleared by informants in 1986–7 only (ha)	109	24	163	296
– As a % of declared plot area under forest cover originally	1.7	3.7	2.0	1.9
– As a % of total declared forest cover area cleared	5.3	4.4	18.2	8.4
– As a % of declared forest cover area cleared by informants until Feb. 1987	8.4	12.0	36.8	15.2
Theoretical clearings by all informants until Feb. 1987 (ha)	305	134	94	533
Average area of forest cover cleared per informant and theoretical clearing by Feb. 1987 (ha)	4.3	1.5	4.6	3.7

Sources: Fieldwork and interviews with 50 plot occupants: 7–28 February 1987 and 17 August–16 September 1987.
Notes: * Includes areas in fallow, in planted pastures and in crops.
    † Area of original forest cover likely to have been appraised less reliably at Vila Brasil, considering the long history of settlement in this region and current practice, observed in many cases, of re-utilising lush regrowth and *cerrado* woodlands.

factors that induce an expansion of clearance activities, of which the more important are:

1 Subdivision of original tracts of land into smaller plots. Many informants expect to be granted property titles for land tracts greater in size than

Table 5.2 Years required for total forest clearance on claimed plots, based on mean rates and areas cleared by informants since arrival and in 1986–87, northern Roraima, Brazil

	Vila Brasil	Taiano	Tepequém	Total
Declared plot area under forest cover originally (ha)	6,603.27	654.00	7,973.50	15,230.77
Total declared forest cover area cleared by informants and previous occupants until 1986–7 (ha)	2,066.83	544.92	894.84	3,506.59
Forest cover area remaining on informants' plots after 1986–7 (ha)	4,436.44	109.08	7,078.66	11,624.18
Total forest cover area cleared by informants in 1986–7 (ha)	109.37	24.14	162.51	296.02
Number of informants	16	12	22	50
Years required for total clearance of forest cover areas	40.56	4.52	43.56	39.27
Average forest cover area cleared per informant per years until 1986–7 (ha)	4.29	1.45	4.63	3.65
Average forest cover area cleared per informant in 1986–7 only (ha)	6.84	2.01	7.39	5.92

Source: Fieldwork and interviews with 50 plot occupants; 7–28 February 1987 and 17 August–16 September 1987.

their likely capacity for exploitation. It can be anticipated that public authorities will only issue legal ownership for much smaller plots of land. Area thresholds at best might approach the regional standard applied in land distribution programmes of the 1970s (i.e. 100 ha), though titling agencies are more likely to follow less generous levels (50 ha and below), as tends to be the case in more recent settlement projects. This granting of ownership to smaller-scale plots than areas claimed should provide a considerable amount of land for new land-seekers. These, in turn, represent many additional units of forest clearance in the region.

2 Increase in the number of workers on each rural estate. Even if the first trend were limited or absent, some of the producer's relations, acquaintances, associates or employees are bound to arrive in the wake of the initial colonisation. These arrivals rapidly enlarge the local supply of labour and the demand for food, thus increasing pressure for further clearance. Also in the longer term, as the household's children reach adulthood and marry, many remain on the parents' property, at least initially, where they establish fairly autonomous farming sub-units exerting further pressure on forest resources.

3 Capital accumulation on plots by the same occupant. As will be discussed later under the heading of social and economic conditions, local people generally tend to invest and accumulate savings in the form of cattle. Any economic progress of the production unit will preferably reflect itself in

an increase in the acreage of planted pasture, a major factor of deforestation (Léna 1988). Savings may begin to appear following a series of good food-crop harvests, when permanent crops come into production or through the rental of planted pasture. Some farmers develop a role in fattening other producers' cattle. Savings also come from earnings in placer-mining or other short-term off-farm employment. These activities include pond and well-digging, pasture-fencing and road-opening.

4 Transfer of ownership to a more successful producer. Some individuals have managed to accumulate sufficient savings, within Roraima or elsewhere, to purchase plots from the more needy settlers. They can be cattle-ranchers, timber traders, merchants, successful *garimpeiros*, well-to-do agriculturalists from southern Brazil, or landowners who sold property in regions with more formalised land tenure and more valuable land markets. Such purchasers usually have more ambitious plans than the seller. There is a need for their investment to generate profits and, in order to do so, they usually have access to larger capital resources. Therefore, even where there is little inward migration, the replacement of occupants through land sales further aggravates forest clearance in areas already settled.

5 Production changes on traditional cattle ranches. In all three localities surveyed, ranchers in the forest–savanna transition zone have been investing in large-scale mechanised agriculture or have plans to do so. Schemes for development often include irrigation in gallery or flood-forest sections of their land, for example the River Amajari margins (Figure 5.1). These tend to be much more productive than the natural grasslands, for example at Vila Brasil.

Several trends increase the extent of annual clearing and the number of farm units per clearing. One concerns the relationships between groups of people involved in driving the frontier of settlement forward. Some of the migrants to Roraima arrive at the forest–savanna transition zone for the locational advantages outlined earlier. Amongst the newcomers, a few ranchers and investors either plan to or have started to acquire extensive tracts of forest in fairly inaccessible areas, and await the opening of roads to turn these tracts into pasturelands. This is the case at the far end of the future Trairão road, which was planned to extend around the foothills of the Serra do Tepequém and open up the still largely undisturbed forest expanses of the upper River Amajari (targeted for sponsored agricultural projects).

As often happens in Amazonia, large tracts of land are being purchased well before roads are built and the infrastructure installed. It is frequently claimed that there is collusion between prospective land purchasers and the local political authorities concerned with land investment interests. Through efficient lobbying, a limited number of investors and large landowners are able to be among the first to benefit from information on the siting and timetable of future development projects. In some instances they may even influence public decision-making (Mougeot 1983: 132).

It may be objected that deforestation will never be total in the localities surveyed. In some marginal zones it might indeed be difficult to convert forest into cropfields on account of flooding, stony soils, escarpments or steep gradients. Such an assumption would be reasonable if mechanised agriculture were to prevail; in this case producers could easily afford to set aside the marginal zones and some degraded areas might even be reforested. The prevailing reality of our study areas however, is very different. Even on the more vulnerable sites, clearance is being practised. For example, at Tepequém, corn is grown and sheep are grazed on very steep gradients with rocky outcrops. Marginal areas in Roraima are unlikely to remain forested given the swift pace of occupation in the future. As indicated by other developing agricultural economies and our own observations in different regions of Amazonia, any tract of forest land that can be used to plant pasture or grow food crops (mainly manioc and corn) is liable to be felled.

In fact, marginal zones are held as fairly fertile by some of the farmers interviewed at Tepequém and Taiano. This belief is part of a well-known 'peasant strategy' in Amazonia which generates satisfactory returns over the short term. Hilltop and foothill soils are avoided, the former usually because they are compacted and impoverished, the latter because they are more sandy and flood-prone. Instead, mid-slope soils are preferred because they are better structured and drained and enriched with nutrients being transported downhill through run-off or mass movement. However, mid-slope soils are eroded as more and more of the forest cover is being cleared. With poor management, repeated cropping becomes disastrous in the longer term. Landless peasants, dependent family members, farm labourers and other rural wage-workers are always strong candidates for marginal-terrain agriculture near or within capitalised farming units. However, for most of Roraima, such use of marginal land is at present small-scale and localised.

Soil conditions[1]

It is well known that the majority of humid tropical soils limit agricultural development (Sanchez 1976; Furley 1990). Differences are also apparent between forest and savanna soils in Brazil (Furley and Ratter 1988) and between evergreen and deciduous forest. The latter is usually associated with better, mesotrophic soils (Ratter *et al.* 1978; Milliken and Ratter 1989).

The three major agricultural settlements lay in approximately similar climatic but different topographic and vegetation units. Consequently a reconnaissance survey was made of soil properties to see whether these also affected the agricultural possibilities and success at each settlement. Surface soils were examined at undisturbed and cleared sites. The properties

Table 5.3 Comparison of soil properties for undisturbed forest and cleared sites at the three study localities

Property	Taiano		Buritizal	Vila Brasil		Maloca	Tepequém	
	Forest	Disturbed		Disturbed	Regrowth		Forest	Disturbed
pH	6.89	4.08	4.37	5.21	6.71	5.82	3.99	4.32
Loss on ignition (%)	17.42	11.28	2.57	3.31	3.18	2.17	11.51	9.37
Organic carbon (%)	15.27	7.36	2.73	5.18	1.91	1.64	9.54	6.54
Available phosphorus (ppm)	1.69	0.69	0.79	1.66	0.90	1.17	0.97	1.31
Exchangeable H (cmol/kg)	2.00	2.60	1.00	0.60	2.40	1.00	5.40	4.20
Exchangeable Mg (cmol/kg)	0.89	0.99	0.60	0.63	0.90	0.38	0.47	0.33
Exchangeable Ca (cmol/kg)	6.72	1.37	0.48	0.92	0.83	4.24	0.25	0.17
Exchangeable Na (cmol/kg)	0.45	0.65	0.52	0.33	0.36	0.03	0.42	0.41
Exchangeable K (cmol/kg)	0.14	0.34	0.17	0.27	0.53	1.17	0.43	0.68
CEC (cmol/kg)	10.60	5.35	2.77	2.75	5.02	5.82	6.97	5.79
Base saturation (%)	77.36	62.62	63.90	78.18	52.19	82.82	22.50	27.46

analysed comprised pH and exchangeable acidity (showing the degree of acidification), loss on ignition and organic carbon (giving an indication of organic matter), available phosphorus, exchangeable calcium, magnesium, sodium and potassium (which offer an insight into the nutrient status). The results are summarised in Table 5.3.

With the exception of Taiano, the soils are generally mildly acidic, with adequate levels of organic matter, low available phosphorus but surprisingly adequate exchangeable cation levels, particularly at the forest sites. Only Vila Brasil showed nutrient cation levels less than the 4 cmol/kg, considered as the threshold for agriculture. Taiano stands out in many respects. It has the highest organic matter levels and pH, high base saturation and adequate CEC figures, dominated by calcium. The established relationship between deciduous forest and mesotrophic soil conditions (even though relatively little is left of the pre-1950s forest cover), seems to be re-confirmed here.

A further feature worth comment is the effect of forest disturbance on soil properties. At Taiano, organic levels have been halved over the period of colonisation, pH has dropped from mildly alkaline to moderately acid, the critical phosphorus levels (taken as below c.10 ppm) have been reduced by over a half and are clearly deficient for all types of agricultural enterprise; the CEC has likewise been halved, although still sufficient for some purposes, and base saturation remains high. Broadly the same pattern is evident with lower overall levels, at Tepequém. There are a few important differences however: pH tends to increase, along with base saturation, exchangeable potassium and phosphorus, when forest is converted to other land uses, but organic matter and most exchangeable cations decline. This may reflect the type of replacement land use but the differences are in any case so slight that it is doubtful whether they are significantly at odds with the general understanding of soil deterioration with forest clearance (Jordan 1987; Furley 1990).

Vila Brasil reveals an intriguing pattern. The results indicate an increase in pH from *buritizal* to disturbed savanna to secondary re-growth, which resembled a nearby *maloca* with sedentary Indian agriculture. Apart from base saturation (possibly related to poor drainage), the *buritizal* soils are poor. It is difficult to explain why disturbed and secondary growth sites should have better soils unless it is related to soil improvement (not widespread), recent burning (not evident) or livestock enrichment (which would be concentrated in limited localities). Overall, however, the acidity, low phosphorus and weak nutrient levels would have a long-term deleterious effect on agriculture without measures to correct deficiencies (see also Chapter 4).

PROGRESS OF AGRICULTURAL COLONISATION

Vila Brasil

Preliminary observations in the savanna areas around Vila Brasil show that, traditionally, extensive ranching estates in the transition zone have always sought out forested tracts. These may be in the form of continuous expanses of high forest, curtains of gallery forest or hill-crest patches of woodland. This has been in order to benefit from the two complementary ecosystems. Annual crops were and still are planted at the expense of cleared and burnt forest. As a consequence of estates having been parcelled out successively from legacies and land transactions, a new kind of ranch has appeared today which lacks any forest remnant. The owner of such a ranch usually tries to come to an agreement with a better-endowed neighbour, whereby he is granted access to a forest patch in order to plant his subsistence crops in exchange for sharing part of the harvest. However, forest-less ranchers rarely manage to produce sufficient corn or banana forage supplement for their cattle, which were observed to be affected by loss of weight and mortality in the dry season.

In recent years, ranches still possessing some expanse of woodland have been securing market values well above those of ranches which lack such resources. It is also the case that on many traditional estates, herds cannot expand without a proportionate enlargement in the area of pasture. This dilemma is confronted by many ranchers who see a pressing need to increase their cash income in order to increase their standard of living. If herds were allowed to multiply spontaneously, stock rates might increase close to or in excess of the carrying capacity of natural grasslands and marginal planted pastures, thus raising mortality rates. This has been reported by several ranchers during the field survey. Now constrained by a land valuation process which impedes most ranchers from expanding their pastures, the ranching economy seeks to survive by developing land areas outside the traditional and largely savanna-dependent system.

Forest expanses are the only areas where land titles can still be secured fairly easily, where yield productivity is greater per area unit and where the return on investment can be obtained rapidly (see section on animal husbandry, pp. 138–41). As a result, there is now a frontier of planted pasture clearly advancing on the forest and away from the traditional cattle estates of the savanna. In the localities studied, this front develops either through purchase or takeover (*posse*) of a tract of virgin forest land. Along the road to Tepequém, this occurs through the purchase of one or more partly cleared plots. Such moves are an attempt to provide a cattle-fattening appendix for forest-deficient ranches of the savanna. Cattle set aside to be sold, post-parturient cows or underweight/weakened cattle can spend some time on the newly planted pastures. Otherwise, according to some ranchers,

Table 5.4 Land use in hectares and as percentage of cleared forest cover area, northern Roraima: 1986–88

	Vila Brasil	Taiano	Tepequém	Total
Plot area claimed by informant (ha)	21,872.09	668.00	9,141.50	31,681.59
Unplanted grasslands and *cerrados*	15,268.82	14.00	1,168.00	16,450.82
Forest cover	6,603.27	654.00	7,973.50	15,230.77
Declared forest cover areas cleared by informants and previous occupants	2,066.83	544.92	894.84	3,506.59
	(100.0%)	(100.0%)	(100.0%)	(100.0%)
Fallow and bush regrowth	1,361.89	359.52	318.37	2,039.78
	(65.9%)	(66.0%)	(35.6%)	(58.2%)
Planted pastures	623.09	114.18	415.01	1,152.28
	(30.1%)	(21.0%)	(46.4%)	(32.9%)
Permanent crop fields	0.80	0.80	5.70	7.30
	(0.0%)	(0.0%)	(0.6%)	(0.2%)
Seasonal crop fields	78.82	51.45	98.82	229.09
	(3.8%)	(9.4%)	(11.0%)	(6.5%)

Sources: Fieldwork and interviews with 50 plot occupants, 7–28 February 1987 and 17 August–16 September 1987.

Notes: 1 The margin of error (total cleared area minus area under land use not exactly equalling zero) is small in all three localities and is due largely to clearings occupied by buildings, gardens, orchards, roads and alleys, streams and waterbodies.
2 The area of forest clearings at Taiano and Vila Brasil probably has been over-reported, due to confusion in some informants as to how to differentiate forest from very old fallow or very dense *cerrado*.

cattle-trading practices in the region would not be viable. With access to planted pastures, one-year old calves can be purchased, fattened for one year and then re-sold with a good profit. New ranches encompassing still uncleared tracts of high forest account for the large share of the area still under forest cover on the average Vila Brasil ranch (Table 5.4).

Forest clearance by ranchers is proceeding rapidly in the area. South-west of Vila Brasil, at the confluence of the Furo Santa Rosa River arm and its Pau Baru tributary, one producer bought tenure rights to a 2,000 ha tract in 1981, of which 1,600 ha were then under forest cover. In fact, only 20 ha had been cleared prior to his arrival, which were then under bush regrowth. Since his arrival, and over a period of only six years, he has cleared 220 ha, where he has planted pasture (*Brachiaria humidicola*) at an average rate of 36 ha per year. A neighbour bought 3,600 ha in 1982 (of which only 6 ha were then under regrowth); five years later he had already converted 100 ha into planted pasture.

These ranchers do not consider that they have any option other than to fell the forest, as they are still going through the pioneer phase of establishing the ranch. In a few years from now, however, when pasture acreage will have expanded up to their expectations and working capacity, they will face one of two alternatives. Either they will have to clear more forest in order to make up for the declining productivity of existing pastures (invaded by natural regrowth and affected by fertility decline), thereby entering a cycle of long-term fallows, or they could opt for the more expensive possibility of tilling, fertilising and replanting the degraded pastures (forming short-term fallows). Each alternative would exert a quite different amount of pressure on the original forest. When queried on which would be their more likely decision, ranchers said they would prefer the first alternative, since it is more land-demanding and less laborious, with logging enabling the producer to recover part of his clearance costs. Nevertheless, in one case a tractor was being used to remove an approximately five-year old regrowth hindering the enlargement of grazing and this might herald a new practice. In any case, and for the time being, the common practice still remains intensively destructive. Riverside clearances in the vicinity of the Maracá Island Ecological Reserve were observed to disregard forestry code regulations on protection of *varzéas* and to involve the control of natural regrowth with applications of toxic herbicides (mostly Tordon 3500).

The forest islands on savanna estates have remained protected until very recently. Older settlers on *fazendas* São Domingos, Liberdade, Morava Nova, south of Fazenda California and in Maloca Aningal, called our attention to the fact that on some woodland fringes and over long periods of time, caimbé trees (*Curatella americana*) tended to invade natural grasslands. These trees eventually multiplied and developed a dense and closed canopy, under which forest species managed to germinate, grow and finally replace the caimbé cover. There could be a slow but gradual natural expansion of the forest over the savanna, this being heralded by colonising species more tolerant of savanna constraints. Other evidence from the Maracá Rain Forest Project research in the region of Maracá Island, points to a natural tendency for forest communities to colonise at least part of the savanna zone over time (Furley and Ratter 1990; Milliken and Ratter 1989). Much of today's treeless landscape of natural grasslands in the Rio Branco region might be explained by long-term clearance and burning of forest in an effort to improve local grazing conditions (IBGE 1981).

There are many services provided by forest areas to the settlers in the localities surveyed. In addition to supporting annual foodcrops, followed by long fallows, they supply settlers with water, fuelwood, lumber and game, as well as forage supplement and protection for cattle against the sun. Currently they are under threat and the less extensive tracts of woodlands are those facing total clearance in the short term. Many producers have already cleared the legally permitted half of the forest cover on their estates. Most ranchers do not allow forest regrowth; instead, they clear annually and plant grasses along with subsistence crops so that pastures will be ready to be grazed by harvest time.

If, as we are led to think, this practice becomes widespread, the ecology of the region will be substantially altered. Forests are important regulators of small-stream flows. A number of them have already become intermittent in the lifetime of some informants. This trend has been increasing according to numerous reports from interviewees and could be signalled by rows of dead buriti palms (*Mauritia flexuosa*). The buriti grow in wet soils and develop along rows indicating permanent or seasonal streamlines. In up-stream sections trees are more sparse in many cases, but they have been reported to be disappearing throughout the area. This suggests that the seasonal water regeneration is becoming insufficient or that ground water is now located at too great a depth to support buriti growth.

Taiano

Taiano provides a good example of the current land-use practices and consequences faced by exploiting extensive forest islands in the region. The forest was totally parcelled out in lots of 30 ha each some thirty years ago, through a state-sponsored agricultural scheme. It has now been almost totally cleared (82.8 per cent). Remnants are only found today on a few escarpments or at the far-end of roadside plots, abandoned due to repeated rice-crop failures as a result of successive droughts in the early 1980s (Lima 1984). In one residual forested patch, species such as biorana, paparauba, angelim (*Hymenolobrium petraeum*) and pau rainha (*Centrolobium paraense*) were common; individuals of *Brosimum rubescens* when measured at 1.5 m height from ground level, showed circumferences of 170, 169, 230 and 527 cm respectively.

The cleared landscape has been largely succeeded by fallow – dense natural regrowth (*capoeirão*) of variable age (Figure 5.2). For many years small plot sizes have been forcing producers to incorporate natural regrowth into the agricultural cycle. During 1986–7 for instance, only 37 per cent of clearings were made at the expense of the forest.

However, this average figure conceals substantial variations from plot to plot, depending on the producer's technical resources. Most small farmers who work their plots manually depend almost entirely on forest clearances. The few with access to mechanisation generally recover fallows under natural regrowth. According to the small landholders, rice growth is hindered on bush fallow. Swift weed invasion is a serious and more extensive problem. This explains the strong tendency to grow corn or manioc on regrowths and rice crops in forest clearings, even if this means planting on clearings of a relative or a neighbour and sharing the harvest.

Dense and high natural regrowth on small plots has clearly constrained the agricultural development of Taiano. A great deal of labour is needed to convert these fallows into cropfields, as opposed to clearing a comparable area of forest. In fact, only in Taiano did informants insist on distinguishing the *capoeira* (i.e. fallow with some secondary regrowth) from the practically

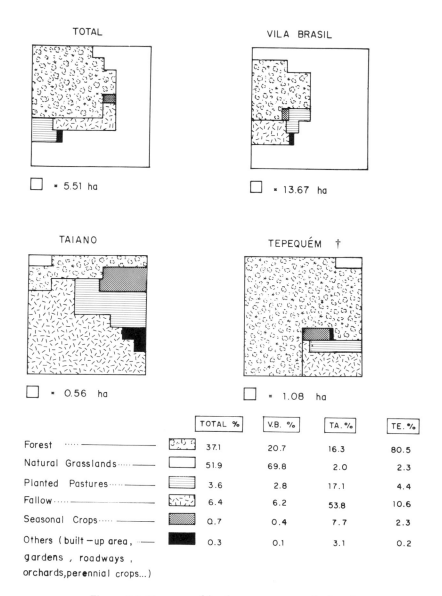

	TOTAL %	V.B. %	TA. %	TE. %
Forest	37.1	20.7	16.3	80.5
Natural Grasslands	51.9	69.8	2.0	2.3
Planted Pastures	3.6	2.8	17.1	4.4
Fallow	6.4	6.2	53.8	10.6
Seasonal Crops	0.7	0.4	7.7	2.3
Others (built-up area, gardens, roadways, orchards, perennial crops...)	0.3	0.1	3.1	0.2

Figure 5.2 Patterns of land use on surveyed plots.*

Source: Fieldwork by the authors in Vila Brasil, Taiano and Tepequém, Roraima, Brazil, 7–28 February 1987 and 7 August–16 September 1987.

Notes: * Average lot size is 551 ha for all localities, and 1,367 ha, 56 ha and 108 ha at Vila Brasil, Taiano and Tepequém, respectively.

† At Tepequém 7 out of 28 lots, with areas over 200 ha, were excluded from computations due to claimed areas being much larger than those of remaining lots and also most unlikely to be titled as such.

impenetrable *capoeirão*. Crops grown in the latter require three weedings during the growing season (at planting, in June and in late August). Only mechanisation could solve the problem and few cultivators were seen or known to have accumulated sufficient savings to reach this stage. Most have to apply for governmental assistance schemes which are currently tied to grain-production programmes.

Many informants observed that the short rainy season, centred on November–December, has tended to disappear in recent years. The main rainy season has grown less pronounced since the early 1980s. This hinders the ripening of seasonal crops (rice, corn, beans). In addition to these reported recent climatic changes, there are indications that extensive clearings might have begun to induce local environmental changes. Manual clearance of the last patches of original forest and mechanised recovery of fallows together with increasing water use could be contributing to a depression of the underground water table, in this way aggravating or extending seasonal water deficits or natural periodic droughts. Despite this, parts of the area are still flooded during the wet season.

This tends to limit the productive use of areas already deforested and to increase pressure on the few remaining naturally wet areas (forested hilltops, streamlines and palm *veredas*). Cropfield invasions by cattle, fenced-pasture leasing by small farmers (or plans to lease) and selective logging for construction materials, signal the growing pressures of ranching on hilltop woodlands. Individuals reported that wells have dried up and all have observed that formerly perennial streams and water holes near *veredas* have become intermittent in the dry season. In some cases, as in February 1987, producers have needed to increase well depths by 2 or 3 m in order to collect enough water to meet their domestic needs.

Clearance has probably contributed to the successive droughts experienced in the early 1980s, mainly through aggravating the intermittent nature of streamflows and seasonal water deficits. According to meteorological data from the Boa Vista Station, during the first half of the 1980s minimum and maximum average monthly temperatures rose, total precipitation fell and rainy seasons became more erratic (Roraima, Seplan-RR n.d.). In any event, we witnessed the difficulties faced by farmers in watering their gardens, cattle and seedling nurseries. Increasing activity has been conducted in the moister valley-bottom lands since early settlement. Because of growing dry-season water shortages and because cleared watersheds have an increasing effect, pressure is now exerted on the lower-lying flats. Here ponds have been dug for the cattle, stream beds have been raised and converted into gardens (at times polluted by nearby pig pens) to make watering more convenient. Producers who do not possess moist areas make arrangements with neighbours to reclaim streams in order to plant more drought-vulnerable crops such as sugar-cane.

According to the Indian chief of the local Maloca Barata, continued forest clearance at the headwaters of the reserve's main stream, the Amorim, over the last 35 years, has caused the flows in this substantial river to become increasingly intermittent. A smaller stream, the Guariba, has not been affected yet but is threatened by the establishment of a cattle estate at its source. This longtime resident has seen forest expanses and *veredas* shrink in size and the yields from hunting and fishing diminish dramatically over one generation.

Technical reports from state and local agencies have been repeating for decades that water deficits are the main constraints limiting agricultural development at Taiano. However, no solution will be viable without major reforestation and a forest protection component. Only the greater use of non-destructive extractive resources and high market-value fruit and vegetable crops with the support of motor water pumps, would significantly improve the well-being of most producers and in turn benefit the Boa Vista market. Such a strategy would generate more attractive prices than rice and be less demanding of space. Nevertheless, such a policy would encourage more intensive forms of cropping on lower-lying areas. The system would be viable in the long term only under certain conditions:

1 leaving natural regrowth to recolonise hilltops and upstream sections of water streams in order to normalise year-round flows (torrential rains and temporary flooding would otherwise damage low-lying crops);
2 expanding rural electrification and providing loans for the purchase of water pumps;
3 providing appropriate storage, grading and transportation for good-quality produce to reach urban markets at fair prices with minimum loss.

Tepequém

The region examined extended along the length of the RR 203, the Vila Brasil–Tepequém road, from the forest–savanna boundary to the summit of the perched syncline where the old diamond *garimpo* of Vila de Tepequém is situated. It is possible to distinguish three types of landscape: the piedmont, which becomes more and more accidented as the Serra is approached; the escarpment and immediate footslope zone; and, finally, the summit of the syncline. The last type is made up of a shallow valley linked to numerous small tributary basins. Savanna predominates over the upland, but tracts of forest encroach upon the scarp zones and along the tributary streams of the summit area.

The settlement represents a pioneer mining community set up in 1937. At the close of the 1950s, the population had reached 4,000 against only 650 in 1980 according to Cruz (1980). The same author also states that the peak of diamond production, reached in the early 1940s, dropped to its lowest after 1960 but recovered later when, in 1977, reclamation was introduced.

Amongst those interviewed, six had worked on the *garimpo* at an early period (1938–1965), abandoned the site and then returned towards the end of the 1970s. The objective of the return was, however, not principally for mining but for agriculture. A report from SEAGRI (Roraima 1979) noted the existence at this time of a small number of maize producers who were reported to obtain twice the production of those in Taiano. Thirteen colonists have arrived since 1980 and in general they have taken the abandoned plots.

With improvements to the road in 1985, in advance of the planned Colonisation Project of Tepequém, there was a fresh wave of settlement. For the most part, this consisted of substitution of new arrivals for older settlers selling their plots. At the same date, it is worth noting a similar thrust of pioneer farmers (cf. Vila Brasil) who bought the 100 ha plots of earlier small-scale colonists in order to create cattle ranches.

Moreover, the development of ranches entailed the invasion of cultivated fields by livestock and many of the smaller agriculturalists were driven to sell their lands. Small-scale cultivators were not able to protect their fields for more than two years. At the same time, the ranchers did not wish to assume the costs of enclosure and protection that they would have to quickly replace as their grazing areas changed. Therein lay the conflict between cultivator and rancher.

The intensity and types of land-use change and the consequences of bringing new land into production have varied with the nature of the zone under consideration: the summit of the syncline, the escarpment and the pediment zone, or the piedmont.

The summit of the syncline

The landscape is essentially one of savanna. It is probable, however, that the forest occupied a more extensive area in the past. It is possible to find fragments of very hard wood in the soil, still resistant to fire and weathering, that the inhabitants gather and use for enclosures. However, given the large number of inhabitants during the years 1940–50 and the absence of access routes for transportation, the miners exercised a great pressure on the wooded zones in order to gain sufficient cultivated land for an adequate food supply.

Today they are using the last fragments of woodland in order to plant their annual crops. The intensity of land use does not permit natural regeneration, condemning the forest to rapid destruction. The forest cover has dwindled to edges and is already insufficient to sustain the livestock of the local people – particularly those who, as a result of a lack of other job opportunities apart from mining, are forced to turn to agriculture. This last group of people continue to live on the summit but are instrumental in clearing forest towards the piedmont zone. They are angered by the impossibility of obtaining a definite title to their land in the mining area as well as a lack of opportunity in the development plans so far announced (routeways and proposed colonisation projects).

The escarpment and the pediment zone

The steep slopes make access very difficult. The single road and few tracks provide the only means of entry and reaching hunting areas. However, observations indicate widespread recent deforestation where fire has spread rapidly. A few small cultivated fields border onto stretches of marsh which have been abandoned and have been left as very poor pasture. A small group of dwellings is also sited on a rocky scarp in order to graze sheep. This is an example, amongst many others, of a very extensive utilisation of marginal zones which would normally remain wooded. If this scale of deforestation is continued, the effects of torrential erosion will be accentuated, aggravating what is already a very precarious situation. The access road is practically impassable in the rainy season and suffers from frequent gullying.

The piedmont

If agriculture is the main occupation in the piedmont zone, the presence of the *garimpo* close by presents an attractive additional opportunity for a number of colonists. There is the possibility of supplementing incomes, and many colonists and their children work in the *garimpo* for part of the year. Furthermore, the activities of the mining group provide a market for the colonists, especially for manioc flour (*farinha*) and poultry which fetch better prices here than in Boa Vista.

The piedmont of the Serra de Tepequém has all the characteristics typical of a pioneer zone:

1 Extensive land use, which attempts to clear the maximum area since this is considered in the assessment of applications for land titles. One example of such a calculation produced an area 5 times that cleared. Such deforestation augments production by optimising utilisation of family labour in conditions of relatively high productivity. The potential of forested areas for rapid clearance is considered as an added value which aids re-sale at a higher price.
2 Considerable turnover of colonist families, which results in part from the difficulties described above and in part from the need to realise capital – provided by the 'added value' of clearance.
3 Weak integration between the different stages of the agricultural cycle. Deforestation occurs preferentially depending upon the type of forest. At Tepequém, 84 per cent of the clearance between 1986–87 was carried out at the expense of primary rather than regrowth forest.

AGRO-PASTORALISM: YIELDS AND LABOUR PRODUCTIVITY

Subsistence crops

These are grown in areas of cleared forest following a general polycultural model but with considerable variation as a result of individual strategies and location.

The basic model consists of a rice–maize–beans and manioc association within the same area, but the size of the tract to be cultivated is an important factor in determining the cropping system. Most often rice is the first crop to be planted in a sequential system, followed by maize, and when that is already established, beans and manioc. There are also interplanted systems of maize and beans, a line of manioc, further maize and beans and so on (see Table 6.2). Rice occupies the remaining area, often on damper patches. Each farmer selects his spacing and arrangement according to his objectives, his past experience, from his estimation of soil quality, and from what he knows of past yields. Where a part of the annual forest clearance includes old natural regrowth, the farmer most often will plant a monoculture of maize, eliminating rice. In previously undisturbed forest, rice is preferred. Perennials such as cocoa and coffee (where there is a market) or bananas, are grown in forest clearings. In contrast, pasture tends to be indiscriminately established on any plot of land. The option of pasture eliminates secondary crops; one would thus have the sequence rice–maize–pasture grass, planted successively in the same area over a period of about one and a half years. Following the rice harvest, then that of maize, all the area is left for pasture.

If the intention of the coloniser is not to create pasture, a whole range of possibilities exist. The basic system can be augmented by the introduction of marrow, sugar-cane, bananas, and perennials (though rare in the region studied). Often the whole field consists of a rice–maize association, but it will be subdivided into secondary associations such as rice–maize–manioc, rice–maize–beans–bananas, rice–maize–marrow, etc.

This complexity makes it very difficult to interpret regional statistics concerning yields of each crop. Unsurprisingly, there was considerable disparity between figures from yields obtained by means of questionnaires and from available regional data. The only method which would give reliable results would be that of annual yields from specific sample plots, which was not possible within the survey period. Even given this detail, the extrapolation would have been a problem in view of the complexity of crop associations in time and space; for example, information gathered orally and from memory is imprecise; the harvest is staggered over time; manioc is not always sold in its raw form and is used in a variety of ways for humans and animals.

130

Table 5.5 Main seasonal crops according to area planted, yield harvested and yield harvested per hectare, northern Roraima: 1986–87

	Number of producers	Producers with data	Area planted (ha)	Yield harvested (kg)	Yield harvested per hectare	
					Average (kg)	Maximum (kg)
Rice						
Vila Brasil	8	8	63.04	36,840	584	1,455
Taiano	7	7	20.22	5,220	258	1,515
Tepequém	18	18	41.35	25,800	624	652
All	33	33	124.61	67,860	545	1,515
Maize						
Vila Brasil	15	14	86.70	76,170	878	2,030
Taiano	12	12	51.61	41,280	800	2,182
Tepequém	20	13	74.84	33,120	443	1,440
All	47	39	213.15	150,570	706	2,182
Beans						
Vila Brasil	9	5	3.52	360	102	750
Taiano	4	3	1.15	320	287	364
Tepequém	9	3	1.58	290	184	240
All	22	11	6.25	970	155	750
Manioc						
Vila Brasil	11	2	3.30	4,200	1,273	1,818
Taiano	7	3	6.93	3,585	517	3,818
Tepequém	16	1	1.32	1,050	795	795
All	34	6	11.55	8,835	862	3,818

Sources: Fieldwork and interviews with 50 plot occupants, 7–28 February 1987 and 17 August–16 September 1987.

Maize accounts for the majority of area planted at each site (Table 5.5). It is second to manioc in terms of return at harvest. It is the crop most often encountered at smallholdings visited, with area, production and net return greater than that of rice. The total yield of the study plots (706 kg/ha/yr) for 1986–87 compares favourably with that for Roraima (596) but is less than that for the Northern Region as a whole (1,328). However, given the best conditions at each location, some practices obtained yields equivalent to two or three times the average for Roraima. The maize yields for Tepequém are poorer, whereas those of rice, usually restricted to cleared forest, are higher than at the other sites.

Overall, rice yields are barely a third of the state or regional average. After having been developed during the late 1970s, the area planted to rice dropped rapidly at the start of the 1980s due to a series of droughts. Since then, and until 1987, the yield doubled in spite of little expansion of area, reaching the yield level of the late 1970s (comparable to regional and national averages). This appears to result from the increasing concentration of production on several large irrigated farms and a reduction in the number of fields on small landholdings.

Despite the relatively poor yields of rice in the study sites, some favoured plots are able to yield three (Vila Brasil) or six (Taiano) times the local average, nearing the average levels for the state. This is particularly the case for the Taiano operations, supported by government credit programmes, technical aid and mechanisation, so that it has become a source of grain for the local urban market.

Whereas maize is grown to advantage on fallows, rice does less well in them as it is less tolerant of lack of water and, in addition, requires much more work at harvest. Seedlings succumb more readily to grass invasion and natural herbaceous and woody regrowth. Consequently, rice is a crop of initial clearance, whilst maize is more versatile.

Compared to rice and maize, the area given over to beans and manioc is minimal, although there is considerable difficulty in assessing yields. The bean harvest for all three sites (155 kg ha^{-1}yr^{-1}) in 1986–87, was about half the state average (293) and much less than the regional value (583). This occurred despite a number of more successful farmers who declared net yields greater than the state average – at Taiano (364 kg ha^{-1}yr^{-1}) and 750 around Vila Brasil. The harvest of beans is more onerous than that of manioc, and it is planted in smaller tracts where production often fails for lack of water. Most of the harvest is for house consumption.

The average yield for manioc at the 3 sites (765 kg ha^{-1}yr^{-1}) is only a fraction of the state average (7,456) and of the region (12,911). Even in the best conditions found at Taiano, individual yields are no more than 3,818 kg ha^{-1}yr^{-1}. These large differences for manioc between census data and the present study, even accounting for the non-use of 'yield quadrants' in our method, can also be largely explained by the widespread practice of planting manioc at different times of the year and harvesting it over a longer period than other crops. This affects the precision of the figures given by farmers and is especially true for manioc. The time to maturity for manioc is one to two years but it can be left in the ground without any serious loss of quality. This allows it to be harvested and converted to flour when labour demand from other activities has slackened in the dry season.

For these reasons rice was selected for a study of labour productivity. Rice is the staple food and is the object of substantial commercialisation (50 per cent on average). The characteristic cultivation activities are concentrated in time and thus readily comparable. Generally, labour productivity is good whereas, as we have seen, yields are low. These two elements are typical of extensive manual production systems. Some 14 days of work, including clearing, burning and weeding, are sufficient to provide an adult with a yearly supply. Several factors, such as planting, harvesting and consumption, influence the area planted by unit time of labour (Table 5.6).

Table 5.6 Rain-fed rice crops according to labour productivity of planting and harvest and per capita subsistence share of harvest, northern Roraima: 1986–87

	Vila Brasil	Taiano	Tepequém	Total
Planting (informants)	(8)	(7)	(21)	(36)
Planted area (ha)	43.44	20.22*	77.73†	141.39
Worker-days	231.00	90.50	317.00	638.50
Area planted/worker days	0.19	0.22	0.25	0.22
Maximum	0.26	0.37	0.66	0.66
Harvest (informants)	(7)	(7)	(12)	(26)
Weight (kg)	19,200	4,860	13,140	37,200
Worker-days	289	234	188	711
Weight harvested/worker days	66.40	20.77	69.90	52.30
Maximum	89.10	90.00	192.00	190.00
Consumption (informants)	(5)	(5)	(13)	(23)
Weight kept for subsistence (kg)‡	8,640	4,140	7,740	20,520
Household members	46	45	61	152
Weight available/household members	187.80§	92.00	126.90	135
Maximum	305.70	136.40	360.00	360.00

Sources: Fieldwork and interviews with 50 plot occupants, 7–28 February 1987 and 17 August–16 September 1987.

Notes: * The relatively small area planted in rice at Taiano could be due to corn cropping being more suited than rice or fallows with regrowth.

† Area under crop in 1987–88 was added in order to base calculations on a reasonable set of data.

‡ Part of which is likely to have been or could be exchanged or sold to buy goods and/or pay workers, etc.

§ Good share of which is probably used to feed temporary workers, whose incidence in total manpower is particularly high at harvest in Villa Brasil, although these are not considered as members of the households.

Planting

The best performances at Taiano included improved labour efficiency through mechanised ground preparation. At Vila Brasil on the other hand, the need to economise on space among forest stumps leads to more crowded planting. Conversely, conversion of the plot to pasture, frequent at Vila Brasil, eliminates secondary crop associations and allows a greater number of plants per hectare. At Tepequém there is greatest allocation of space and, unlike the other locations, space is not a limiting factor. On the contrary, as much as possible is used. Here there is an increase in the predominance of fields under associated crops. Nearly all the clearing is done at the cost of primary forest. The first burning leaves an important amount of unburnt organic matter. The clearings intended for pasture, opportunistic or low-intensity, are often of poorer quality and are not as intensive as rice plantation.

Harvesting

The labour efficiency of harvesting at Taiano is very low. Nearly all those questioned confirmed that manually cultivated rice is inconvenient. Because it is grown in fallow areas, encroachment by other grasses is intense and the ear-growth is poor. The farmers attribute the latter to a lack of water, particularly for short-cycle rice. Equally, competition from regrowth should be taken into account. Usually the harvest is performed by the family, assuming that the areas are small. This results in lower labour efficiency (participation of young children and the undertaking of other activities at the same time). The forest clearings, as at Tepequém, give a good index of labour productivity.

Consumption

The per capita consumption at Taiano is scarcely sufficient. Nearly all is consumed at home. A number of people prefer not to depend solely on rice and try other crops (maize, chickens, pigs or manioc). They are free to buy their rice from neighbouring areas which produce a surplus (Alto Alegre, Tepequém). At Vila Brasil, the use of hired labour for a range of tasks involves an increase in the amount of rice kept for domestic use, as the provision of food for labourers is necessary.

As a general rule, an immigrant peasant in Amazonia can only have an improved future if his production increases considerably in terms of area, volume or value, through an accessible market. There are thus two possibilities: (1) increasing current production of crops such as rice or maize by increasing technical inputs; (2) selecting produce of high value for which capital investment would be least (cocoa, coffee, rubber, fruits, vegetables, etc.). These two alternatives allow an increase in labour productivity, but the first would only be compatible in well-drained soils (*terra firme*) with ecologically satisfactory environmental management and both strategies would require better developed storage facilities.

Certainly, peasant farmers can increase commercial quantities of crops in most cases, but they are close to their limits, considering their level of technology. In the savanna–forest transition zone, where the tree cover has been practically eliminated, the sustainability of livelihoods becomes degraded little by little. This is due to a progressive deterioration of natural factors of productivity and an increased scarcity of natural subsistence products.

As a consequence, human effort has to increase and hence the cost of subsistence. In the absence of a sufficient energy input from outside the system, the profit margin diminishes because of the effort and hence the accumulation of capital slows down. This situation, already well established at Taiano (the most capital-intensive of the study sites), threatens to occur in the medium term in areas of recent deforestation, such as Tepequém.

To increase production further, a considerable improvement in the infrastructure (roads, transport, market systems, credits, health, schools, etc.) is required, signifying a higher cost to society. However, the creation of such an infrastructure involves land valuation. It is possible to verify everywhere that the attraction of a rapid gain, either to begin once again elsewhere (but this time furnished with some capital), or to change activities to prevent the sale of land, is extremely strong. As long as there is no land obstruction (real or artificially created by the establishment of forest reserves) and hence a rise in land price, no overall intensification seems possible. Even the buyers of land who replace the original colonisers, resort by preference to ranching.

For the common products (rice, maize, beans), smallholders are in competition with mechanised agriculture. If their production costs are lower (partly as a result of being unquantifiable and associated with poor living conditions), the difference is absorbed by the commercial network, in particular when the product is trucked over long distances. On balance, it seems that this marginal 'peasant' activity at the 'frontier' is not viable.

This form of land occupation is best understood as part of a geopolitical strategy of military origin. It alleviates the pressure on urban peripheries, with the social problems they incur. Nevertheless, it results in relative economic stagnation and ephemerality if, by one of the two alternatives outlined above, an increase in commercial production is unattainable. Finance tends to concentrate on the 'efficient' agriculture of the south of Brazil.

The main opportunity for peasant farmers is the exploitation of traditional Amazonian crops. Yet here again, crops originally from Amazonia have been made a success elsewhere, such as cocoa in Bahia, coffee in São Paulo (and even rubber in favourable conditions). It is thus a vulnerable income. On the other hand, the threat of overproduction and low prices is real, particularly if a system of perennial crops was to extend to the majority of the small producers.

A third way to increase labour productivity, which is both practical and meets fundamental needs, would seem to be ranching. Ecologically unsatisfactory, although its impact can be significantly reduced, it is none the less constantly increasing among smallholders, unlike perennial crops. Our three sites conform to the rule.

Cash crops

These show little diversity, consisting only of coffee and cocoa, and do not figure greatly in the sample. A total of 38 per cent of interviewees have at least some coffee and 16 per cent grow it on a commercial scale. For cocoa the situation is even less developed, the respective figures being 30 per cent and 4 per cent. In most cases they are either trials or for home consumption, not exceeding a few plants. Some smallholders received plants from the government (50 to 100 stems) in order to start a plantation.

Cocoa is most prevalent at Tepequém, where two farmers grow it on a commercial scale and six are starting to develop a plantation, with plants and seeds from an agent of the Agriculture Secretariat in Boa Vista. It is planted in the shade of banana trees and seems to do well. At Taiano and Vila Brasil, cocoa is merely on a trial scale with a few stems. The reasonably fertile soils are good for cocoa, as long as it is planted in shady, humid areas. One of the factors inhibiting development is the lack of an assured market. Coffee, present everywhere on a limited scale, is absent from Taiano even at an initial stage. Four farmers at Vila Brasil and four at Tepequém cultivate coffee commercially. In the case of Vila Brasil it is essentially performed by farmers with land on the Tepequém road. Given the number of plants already in production, this zone is becoming a source of plants and seeds for those wishing to adopt the crop. Lack of plant availability is a frequent complaint. At Taiano the few plants observed seem to suffer during the dry season. At Vila Brasil, the summer drought led to the loss of 3,500 coffee plants through an uncontrolled fire.

In fact the best zone for the development of cash crops appears to be Tepequém. The development of the colonisation project in terms of volume of technical and financial assistance, practical support for permanent crops, etc. will obviously influence future trends, but at present many farmers foresee a considerable extension of their plantation.

Fruit crops

All the producers interviewed have fruit crops on their plots. The number of trees and species diversity varies from one plot to another, with specific site differences. For the three regions overall, the most widespread crops are citrus (orange, lime, sweet lime, tangerine) and pineapple followed by crops more tolerant of poor edaphic conditions and longer dry periods (mango and cashew at Vila Brasil and Taiano), or plants which have extensive root development enabling them to tap the water table (mango, coconut, cashew, papaya).

The highest species diversity occurs in the continuous forest zone at Tepequém (27 species) and forest–savanna zone at Vila Brasil (10 species). This is partly due to better soil conditions and humidity, and to the presence of native forest species.

The inventory of fruit species used by farmers is certainly more diverse, with species occurring at greater densities than is suggested by the given data. Often we were able to establish that the producer had failed to mention particular species for various reasons: either they had not been planted that year or they were disease-free and so did not require any particular attention (guava, papaya, mango). This is above all the case with species characteristic of growth in natural forest (palm fruits, native species such as inga, biriba, jabuticaba, etc.).

The lack of water in the dry season, particularly for young plants, hampers the fruiting of introduced species. Additional constraints include the variety and abundance of ants and insect pests which cause damage to stems, foliage and fruits.

These problems particularly afflict orange trees. Only the good market prices, locally and regionally, provide an explanation for their cultivation. The cultivation of orange and lime has undergone a distinct expansion in Roraima during the late 1980s in terms of area harvested, volume and value of production. From 1984–86 the area of orchard increased by 12 per cent in Brasil as a whole and 335 per cent in the northern region. In Roraima, it grew from 92 to 363 ha (294.6 per cent increase). In contrast with national and regional patterns and in spite of poor yields (gains in volume produced were only a third of the increase in area cultivated), the value of production grew to four times or twice the regional levels. This was in part attributable to the number of young orange trees brought into production.

Roughly a third of those questioned apply protective chemicals, particularly to oranges (Aldrin, Nitrosin, Mirex, DDT), but also to avocado and coconut palm (Aldrin), to cashews (Malatol) and other fruit trees (Nitrosin, Actiban and Permosi). Several of these products have been banned from sale elsewhere for years due to their high toxicity (Zambrone 1988). Yet they are also used by farmers in their vegetable gardens, especially on tomatoes (DDT, Baygon, Nevapol, Nirosin, Actiban, Permosi, Malatol) and on peppers (Aldrin).

Home gardens

Nearly all the farmers questioned have at least one *canteiro* (small enclosure of ground, or wooden frame on piles to protect against parasites and poultry), where herbs are cultivated for cooking and medicinal plants are grown. Some householders plant a larger quantity or a more diverse range including tomatoes, peppers, lettuces, cucumbers, cabbages and shallots. Very few, however, reach a commercial scale. The greatest obstacle is the organisation of transport to selling points and other problems include lack of water and pests. Planting occurs in the dry season because it partly avoids the problem of pests but has the disadvantage of requiring watering. If planting is carried out during the rainy season, chemical applications are considered necessary. Kitchen gardens represent the only form of intensive agriculture in the sample, with domestic waste and animal manure being used as fertiliser. However the scale is tiny (some *canteiros* are less than 2 × 2 m).

At Taiano, due to the lorry that takes produce to the Boa Vista market once a week, the transport problem can be regulated (though only partly, as the quantities allowed for each producer are modest). The principal problem here is that of water. Even those who have a well or waterlogged area (typically a *buritizal*) must pump water two or three metres in summer. This is a considerable task and limits the scale of production. It would require use of a motor pump, but the absence of electricity and suitable credits have discouraged attempts.

At Vila Brasil some people use a diesel pump[2] but do not sell much because they lack their own vehicle and have no access to organised transport; another reason for lack of sales is the distance from production to market. Kitchen gardens are situated close to waterholes or by streams, but the latter tend to dry up in summer. Two smallholders interviewed have the use of a vehicle and a source of permanent water and intend to commercialise their production.

At Tepequém, which lacks an irrigation system, most smallholders plant their *canteiro* during the rainy season. Some plant by the river and can thus water the vegetables, but this again involves considerable labour. Two farmers from the piedmont zone sold their produce by horse at the *garimpo*. Eventually they were forced to abandon the scheme because of the small quantities they were able to transport, which made their prices uncompetitive with those higher up on the Serra where produce is forwarded in larger amounts by the lorries coming from Boa Vista.

Animal husbandry

Characteristics and yields

Although only a minority of farmers can be said to own much land, some degree of ranching is widespread in the three areas. It is a strategy for capital accumulation and also protects savings against inflation. Beasts are sold or bartered for manufactured goods and property, to repay debts, to pay for health services and education, for wages or to return favours. In prolonged shortages, they are eaten or exchanged for essential foods. The extent of ranching among the farmers, and the size of their herds, varies from one site to another according to available environmental resources. These include large areas of pasture for the larger animals, or game as supplementary protein in the case of small animals.

For all three sites, ranching of small livestock is most widespread, followed by medium and large animals (Table 5.7). In the large *fazendas* of Vila Brasil, most of the farmers interviewed raise all three categories of animals, and herd size is larger than at Taiano or Tepequém. The extent of ranching, especially that of smaller animals, is greater at Taiano than at Tepequém. However, Tepequém plots are larger and herd size of large

138

Table 5.7 Percentage of informants with livestock and numbers per category of livestock, northern Roraima: 1986–87

	Vila Brasil (16)	Taiano (12)	Tepequém (22)	Total (50)
Average area in planted and unplanted pasture per producer (ha)	993.25	10.68	71.95	352.06
Beef				
Breeders as % of informants	93.80	41.70	36.40	56.00
Average no. December 85	245.58	16.00	37.20	146.00
Average no. December 86	294.57	28.20	72.12	179.33
Pigs				
Breeders as % of informants	100.00	75.00	45.40	70.00
Average no. December 85	30.89	4.20	8.00	17.75
Average no. December 86	25.62	7.67	14.70	17.89
Poultry				
Breeders as % of informants	93.70	100.00	81.80	90.00
Average no. December 85	71.43	63.57	30.89	53.17
Average no. December 86	86.27	45.25	58.50	64.22

Sources: Fieldwork and interviews with 50 plot occupants, 7–28 February 1987 and 17 August–16 September 1987.

animals is greater than at Taiano, where farmyards dominate. This form of animal production is imposed by the scarcity of plots. The strategy at Taiano possibly aims at compensating lower productivity with hunting and fishing. Despite the constraints imposed by the divisioning of plots and the potential conflicts with unfenced crops, group sizes of larger beasts increased notably between 1985 and 1986, largely due to the recent development of pasture planted on consolidated plots. In fact the average densities for medium and large animals estimated for Taiano is now almost equal to those of Tepequém (1.0 and 0.2 head/ha respectively).

Following the trend at Taiano one can predict that ranching will exert a greater (and stronger) influence in agriculture. In 1987, many farmers complained of damaged or wrecked harvests, due to encroachment by pigs or cows from their neighbours. During the same period, whilst farmyards increased at Vila Brasil and Tepequém, they declined at Taiano. Here the long succession of droughts in a zone already short of water, and apparently with less game, forced families to over-exploit farmyards for consumption or sale in order to obtain basic foodstuffs. The available rice per head harvested per production unit is particularly low at Taiano, probably less than the annual minimum requirement.

A comparison of utilisation levels in the three localities clearly shows the importance of natural productivity and the land potential for ranching in the region (Table 5.8). Utilisation level is defined here as the number of livestock sold or consumed during the year, relative to the herd size at the end of that year. For the three (size) categories of animals, the total for the

139

Table 5.8 Number sold and consumed as percentages of end-of-year totals and sales/consumption ratio, per category of livestock and per producer, northern Roraima: 1986–87

	Vila Brasil	Taiano	Tepequém	Total
Beef				
Numbers owned December 86	294.57	28.20	72.12	179.33
Numbers sold in 1986	29.23	0.75	5.00	18.55
Numbers consumed in 1986	5.53	4.00	2.50	4.77
Numbers sold and consumed as % of head owned in Dec. 86	11.80	16.84	10.40	13.00
Numbers sold as % of head consumed in 1986	528.57	18.75	200.00	388.89
Pigs				
Numbers owned December 86	25.62	7.67	14.70	17.89
Numbers sold in 1986	12.07	0.60	8.00	8.54
Numbers consumed in 1986	6.85	2.00	8.40	6.13
Numbers sold and consumed as % of head owned in Dec. 86	73.85	33.90	111.56	82.00
Numbers sold as % of head consumed in 1986	176.20	30.00	95.24	139.31
Poultry				
Numbers owned December 86	86.27	45.25	58.50	64.22
Numbers sold in 1986	18.89	11.33	31.58	22.85
Numbers consumed in 1986	44.50	49.00	44.27	45.69
Numbers sold and consumed as % of head owned in Dec. 86	73.48	133.33	129.66	106.73
Numbers sold as % of head consumed in 1986	42.45	23.12	71.33	50.01

Sources: Fieldwork and interviews with 50 plot occupants, 7–28 February 1987 and 17 August–16 September 1987.

three sites decreases from largest to smallest animal (over 100 per cent or 1.0 for poultry and minimal for cattle).

The level of use remains inversely related to the level of commercialisation (numbers sold/numbers consumed). This indicates that cattle are a capital reserve, eventually used as a source of revenue, whereas chickens are solely a food source for family units (pigs usually meet both needs). Also, the volume of chicken consumption varies only a little from one site to another, independent of the size of the 'farmyard'.

For specific locations, the lowest consumption of cattle at Tepequém could be explained by the fact that herds are undergoing expansion, whilst the proportion of wild meat and consumption of pigs is greater. At Vila Brasil the consumption of pigs is smaller, compensated by greater reliance on cattle.

The situation seems critical at Taiano. Not only are the herds smaller than at the two other sites but the levels of use of cattle and chicken greater. The commercialisation of all animals is also much less than it is at the other locations. Consequently, at Taiano the domestic source of animal protein may be in greater demand because of the smaller plot size, the greater need for human effort, and the low availability of rice and bushmeat. This is likely to hamper capital accumulation, acquisition of plots in more productive environments and the creation of a less labour-intensive and more lucrative system of use.

Factors of production

Livestock farming generally has low technology, as indicated by the levels of feeds, mineral licks, immunisation, artificial insemination, enclosures, pig and hen pens. These production factors are used to varying but low levels in all locations, especially in the case of cattle-rearing, followed by poultry, pigs and horses. Cattle-rearing is characterised by greater use of mineral salts and immunisation, whereas feedstuffs and hen-houses dominate in poultry rearing.

In terms of differences in level between sites, cattle rearing at Tepequém seems to have the greatest technical input. The inadequate sample does not allow definite conclusions, but use of minerals, immunisation, artificial insemination and enclosures is more widespread here than at Vila Brasil and Taiano. This suggests a higher technology in ranching associated with implantation of pasture in forest areas of sufficient size to generate reasonable income. Commercialisation stimulates measures to increase productivity and quality. This also applies to pig-rearing. The greater use of feedstuffs at Taiano for cattle may imply a need for food supplements due to the reduction of stocking capacity in the dry season.

In terms of marketing, the progressive monetarisation of the local economy and the introduction of planted pasture are the source of a recent modification of cattle markets during the year. It is otherwise a very seasonal activity according to local ranchers. This change can be attributed to two trends: firstly, the progressive impoverishment (loss of buying ability) of small farmers whose systems have not developed due to a lack of capital; secondly, the occurrence of new small-scale ranchers, also without sufficient capital, who have to price animals competitively in the short term to obtain revenue and reinvestment.

The first situation consists of farmers who scarcely survive off ranching, who lack woodland to create pasture and capital to pay for labour. In the second situation, the newly arrived ranchers allow their animals to graze unrestricted. They buy and resell cattle rapidly fattened on pasture implanted in woodland. Given that this strategy can be implemented at any point during the year, they can be flexible over selling. The expanding local urban market around Boa Vista exerts a demand felt thoughout the year and contributes to the lessening seasonality of market activities.

The stocking capacity of cattle is four to five times greater on planted than

141

Table 5.9 Mean beef-cattle stocking rates per unplanted and planted hectares of pasture, northern Roraima: 1986–87

	Vila Brasil	Taiano	Tepequém	Total
Pastureland area				
Total	15,891	128	415	16,434
Unplanted	15,269*	14	†	15,283
Planted	623	114	415	1,152
Number				
Total	4,124	141	577	4,842
Adult only	1,956	85	384	2,425
Numbers/ha on planted and unplanted pasture				
Total	0.26	1.10	1.39	0.92
Adult only	0.12	0.66	0.92	0.57
Numbers/ha on unplanted pasture				
Total	0.27	–	–	–
Adult only	0.13	–	–	–
Numbers/ha on planted pasture				
Total	‡	1.23	1.39	1.31
Adult only	‡	0.74	0.92	0.83

Sources: Fieldwork and interviews with 50 plot occupants, 7–28 February 1987 and 17 August–16 September 1987.

Notes: * Total area of unplanted pastureland includes areas of *cerrado* (*caimbezais*) where grass cover is less dense.

† One typical case was excluded from the sample on which Tepequém calculations are based. The producer raised 350 cattle on 1,000 ha of plateau savanna, thus achieving a stocking rate of 0.35 head/ha, higher than figures generally found on low-lying savannas. The difference is likely to be explained by the higher year-round humidity found in the grasslands on the plateau.

‡ Area in planted pasture is still relatively small but the food contribution to the region's otherwise land-intensive system of cattle-raising is significant. Planted pasture makes up only 3.7% of the total pastureland area.

on natural pasture (Table 5.9). Furthermore, cattle gain weight around twice as fast, and more consistently on planted pastures. The advantages are sufficiently clear to explain the drive towards deforestation and the difficulty of reversing the trend. The improvement of natural savanna requires large investments and cannot be conceived without intensification of agricultural activities. For example, it would be ineffective to plant richer fodder grasses on unimproved soils. Equally, the genetic improvement and improved health of the herd will not occur except where they benefit from an adequate diet. It would, in fact, require a complete change of the production system which cannot take place without a transformation at the level of demand and supply, probably allied to limiting forest access.

Similar patterns of ranching are also likely to extend to small-scale agriculturalists who move into forest areas. The difficulties of arable farming, such as seasonal stress, labour problems and irregularity of harvests, make ranching even more attractive. In summary, the greatest advantages of ranching are good protection from inflation, rapid capital growth with money realisable at any moment in case of illness or accidents, excellent labour productivity, low hired-labour requirements, independence regarding harvest times (lack of seasonality allows best prices to be reached), and easy transport (animals eventually get to the point of sale themselves!). Land and equipment valuation is absorbed by the sale of valuable trees and of food crops prior to expansion. Pastures can also be let until the proprietor has the means to buy his own livestock. Conversion to pasture represents an extension of the useful economic life of the cleared area, thus improving the labour productivity of clearance.

In reality there is no model that competes with ranching. Crops are often considered as a means of switching to ranching. At Tepequém the biggest coffee farmer intends to cut back on his plantation in order to start ranching. It is very difficult to go against this logic for it is a strategy that is altogether rational and efficient in the current circumstances. It is not only the modification of these circumstances (product price, demand, transport, land access barriers, aids and credits, etc.) that could lead to a reduction in ranching. The introduction of a new product at a sustained price could alter the situation if the price was appropriate, but would probably not significantly change the trend. In such a situation, the only means of avoiding forest loss in the short term would seem to be the law, by means of forest policy and land controls, despite all the known problems of monitoring and enforcement. In the medium and longer term, revaluation of the forest will be required as a heritage and as a currently unrecognised economic resource.

SEASONAL DIMENSIONS OF NATURAL RESOURCE UTILISATION

The various agricultural activities analysed previously are part of complex systems of resource utilisation. Such multiple systems consist of activities arranged in time and space so as to minimise workload and optimise the well-being of the farming unit. These take into account the unequal distribution of environmental constraints and opportunities, both natural and man-induced.

The previous sections have emphasised spatial variations in explaining major differences in the composition of agricultural systems in three areas. These systems basically occur everywhere, with minor local variations. In forest–savanna zones, the economically more efficient systems are characterised by their adaptability to temporal variations. Such variations include seasonal and annual changes in the types and intensity of environmental stress. This chapter concludes by examining the effects of these constraints on the vulnerability of current systems, in the context of progressive deforestation.

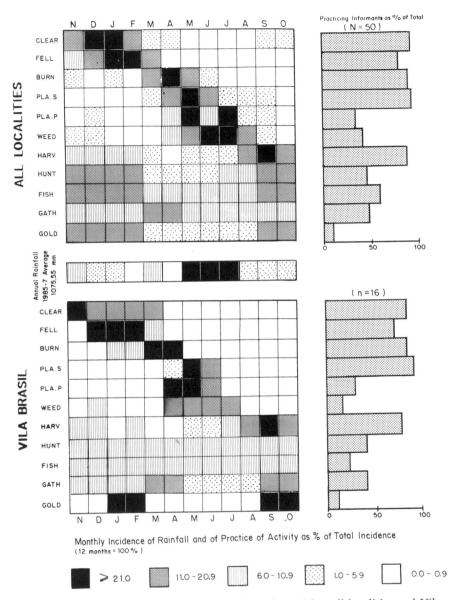

Figure 5.3a Temporal patterns of selected rural activities: all localities and Vila Brasil.

Source: Fieldwork by the authors in Vila Brasil, Taiano and Tepequém, Roraima, Brazil, 7–28 February 1987 and 7 August–16 September 1987. Rainfall data from IBGE, *Anuário Estatístico do Brasil 1987–1988*, Rio de Janeiro, 1988; and IBGE, *Anuário Estatístico do Brasil 1986*, Rio de Janeiro, 1987.

144

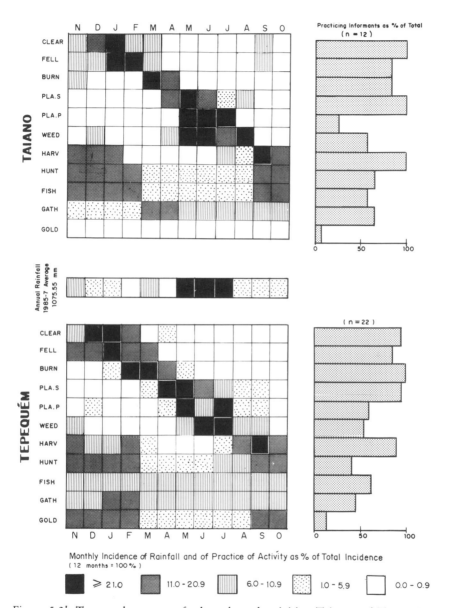

Figure 5.3b Temporal patterns of selected rural activities: Taiano and Tepequém.
Source: Fieldwork by the authors in Vila Brasil, Taiano and Tepequém, Roraima, Brazil, 7–28
February 1987 and 7 August–16 September 1987. Rainfall data from IBGE, *Anuário Estatístico
do Brasil 1987–1988*, Rio de Janeiro, 1988; and IBGE, *Anuário Estatístico do Brasil 1986*, Rio
de Janeiro, 1987.

145

The farmers interviewed in the three areas have developed strategies allowing them to distribute labour through the year to maximise the well-being of their smallholdings and to reduce their vulnerability to adverse environmental conditions. This is illustrated by a comparison of specific activities and their distribution in time, on the one hand, and of the perceived incidence of certain environmental (essentially climatic) constraints and their monthly variation, on the other. Strategies are essentially conditioned by the intensity and seasonality of the rains. The farmer has to judge the potential risks and to reduce the impact of these constraints on the efficiency of his farm.

Generally there is a complementarity between truly agricultural activities and hunting, fishing, gathering and gold-panning (Figures 5.3a, 5.3b). The farmer follows an established agricultural sequence, whilst the other activities are more opportunistic. There is a concentration of the latter activities during land preparation (cutting, clearing, burning) since they are not incompatible. Essentially, agrarian activities are divided between the planting of short-cycle crops and their harvest, and hunting and fishing. In Figure 5.3, monthly figures which are similar or identical between activities or location can hide differences in terms of numbers of producers involved. A high monthly occurrence of an activity is not always synonymous with general practice, as shown by the histograms at the edge of the charts. In other words, an activity can be undertaken over many months by few farmers, or over a few months by the majority of farmers.

The monthly distribution and intensity of agricultural activities are fairly uniform throughout the year in all three localities. This reflects a balanced allocation of labour. Maximum demand is in May, corresponding to the peaks in intensity of planting of short-cycle and perennial crops, the end of burning and the start of weeding. For each of the specified activities, the period is fairly concentrated in time. The maximal incidence of agricultural activities occurs from December to September, the end of one overlapping the start of another.

In contrast to the agricultural cycle, extractivist activities are evenly spread throughout the year (there are no high or zero values in any month). Except for plant extraction, the peak incidences for hunting, fishing and mineral exploitation coincide. This suggests the use of multi-functional areas by the same people, or different areas by several different members of the farming unit.

Generally, the height of extractive practice coincides with the most labour-intensive agricultural activities (burning, seeding, weeding). Moreover it occurs during the dry season, when food and cash resources are at their lowest. Clearance activities also facilitate hunting and fishing. The peak of gold-panning in the dry season ensures that it avoids competition with mechanised extraction (requiring more water) and labour-intensive agriculture. It can, however, hinder expansion of crops and pasture (by deforestation). This is not so much the case for hunting, fishing and gathering, which are either carried out for absolute necessity or can easily be accomplished close to the farm.

The nearest rainfall data are from Boa Vista, in the middle of savanna. According to information from the farmers at the forest–savanna transition on their sequence of activities, the rainy season is more drawn out as one enters the forest zone. This is particularly true of April, when the first rains fall in all three areas. This explains the start of planting in April, even March at Tepequém (largely in forested zone). Generally, planting lasts longer at Tepequém as the threat of lack of rain is less.

The harvest peak for all areas occurs in September, due to the gathering of rice which involves a lot of labour. A relatively intense period of harvest follows for four months. This corresponds to the harvesting of maize (and to some extent manioc), which is more important here than rice yet requires less of a concentration of labour.

At Vila Brasil arable activities are not the most important. They occur in forest and thus have similar characteristics to the other two sites. However, there is a more concentrated planting period due to the limited rains. This is in contrast to the relatively extended harvest, resulting from the importance of maize and manioc as animal feed.

Burning occurs later at Taiano, partly because the area is mainly fallow and the reason is to prevent regrowth before planting. It is also the site where the survey showed weeding to be most important.

At Vila Brasil fishing and hunting do not show any marked seasonality, probably as a result of the proximity of several varied ecosystems and because arable activities do not have the same importance as in other locations. As a ranching area, Vila Brasil can rely on protein from other sources, whereas at Taiano, despite a degraded environment, people rely on wild sources.

Environmental and economic constraints

More than half of all farmers interviewed identified times of the year when certain environmental constraints are more severe (Figures 5.4a, 5.4b). Those which concern the majority of farmers comprise transport difficulties and high merchandise costs because of road conditions (especially at Tepequém), and illness in addition to malaria and 'fever' (especially Vila Brasil). All these constraints are relatively less severe at Taiano. A minority of farmers complained of increasing indebtedness (particularly Taiano), malaria (Vila Brasil), scarcity of food and 'fevers' (Tepequém).

Transport depends essentially on road and bridge conditions. The bridges are constructed of wood and are badly maintained, whilst the roads are dirt tracks and are poorly drained. The access route to the Tepequém plateau is very stony, subject to debris slumps and gullying during rains. Traffic, and

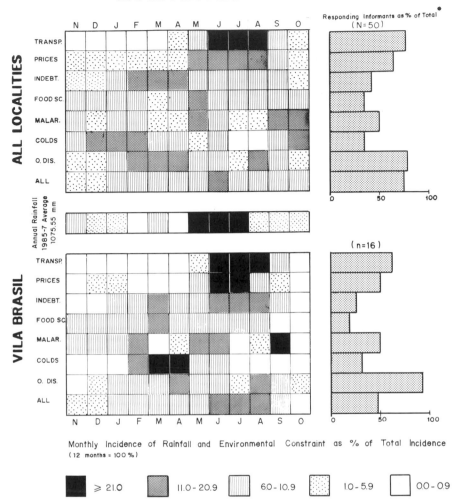

Figure 5.4a Incidence of selected environmental constraints: all localities and Vila Brasil.

Source: Fieldwork by the authors in Vila Brasil, Taiano and Tepequém, Roraima, Brazil, 7–28 February 1987 and 7 August–16 September 1987. Rainfall data from IBGE, *Anuário Estatístico do Brasil 1987–1988*, Rio de Janeiro, 1988; and IBGE, *Anuário Estatístico do Brasil 1986*, Rio de Janeiro, 1987.

Note: * Percentages based on informants able to specify periods of greater incidence only.

hence price of transported merchandise, reflects the state of the road network. Thus the severity and duration of transport constraints vary with relief and force and duration of the rainy season. At Vila Brasil and Taiano, these problems only affect a minority of farmers and occur for only a short torrential rain period with obstruction of savanna routes. On the other hand, at Tepequém, serious problems of transport afflict most farmers over a

148

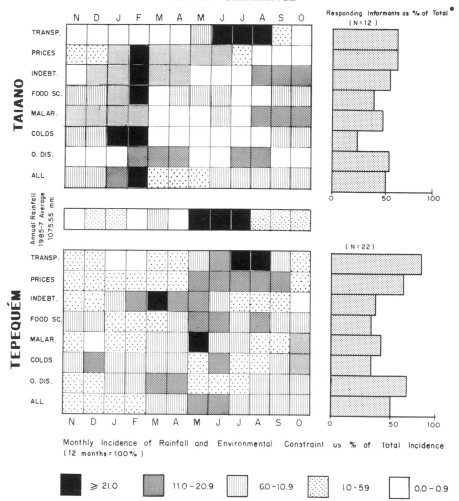

Figure 5.4b Incidence of selected environmental constraints: Taiano and Tepequém.
Source: Fieldwork by the authors in Vila Brasil, Taiano and Tepequém, Roraima, Brazil, 7–28 February 1987 and 7 August–16 September 1987. Rainfall data from IBGE, *Anuário Estatístico do Brasil 1987–1988*, Rio de Janeiro, 1988; and IBGE, *Anuário Estatístico do Brasil 1986*, Rio de Janeiro, 1987.
Note: * Percentages based on informants able to specify periods of greater incidence only.

longer period. This is due to the steep access, longer rainy season, with running water making road maintenance difficult. Such conditions are likely to worsen following deforestation and exposure of slopes.

At Taiano transport problems are less severe and high product prices occur during several months over the dry season. This precedes the period of worst transport. The roads in the undulating terrain around Taiano are better drained but the extent of commercialisation and hence dependence on

149

transport is less than at Vila Brasil or Tepequém. Thus, despite the absence of high prices attributable elsewhere to transport problems, merchandise at Taiano can be considered more expensive. Largely this is a result of a low purchasing power during the distinct and drawn-out dry season. The drop in purchasing power is due to several linked factors: lower productivity, lower potential marketability of harvest crops – hence a poorer economic situation and a greater dependence on food purchases.

Another aspect affecting Taiano is that the major impact of several constraints such as indebtedness, lack of food and illness, tend to coincide with high prices, mostly during January and February. The preparation of land is carried out mainly in these two months, requiring paid and fed labour to keep on top of natural regrowth. This must severely affect a good proportion of the Taiano community. The occurrence of peaks in such impact assessment, suggests a coincidental exposure to the constraints, thus a greater risk of actual reduction in the well-being of the group.

Tepequém and Vila Brasil are not free from these constraints, but they show a less marked seasonality. The peaks of intensity succeed each other and tend to arise later in the dry season (Vila Brasil) or closer to planting (Tepequém). Here, except for transport and price, the stress is better distributed, and can be better managed by the local community. At Tepequém, for example, shortage of money occurs at the same time as labour demand for conversion of forest into fields, yet at a time when the remainder of the harvest is able to meet the farmers' needs. Scarcity can only develop during the rainy season. At Vila Brasil the shortage of food is both less severe and more diffuse than at Tepequém. Maximum indebtedness corresponds to the height of the rains, when few beasts are sold, and to the intervening period between harvests.

CONCLUSION

As outlined in earlier chapters, forest clearance is now proceeding well to the north of the Amazon river. Not only is deforestation affecting the dense forests of southern Roraima, particularly along the major highways, but it has also moved rapidly into the scattered woodlands and forest–savanna transitions which characterise much of the northern part of the state. The Roraima administration has encouraged population growth by publicising land distribution and grain production schemes. The pace of agricultural colonisation by smallholders has accelerated with a commensurate expansion and improvement of the local road network. The traditional commercial livestock operations are struggling to become more efficient in order to compete with products from cattle-rearing areas to the south. All these factors have contributed to forest depletion in a climatically unreliable, ecologically fragile, though apparently economically attractive region.

This chapter has examined the recent evolution, prevailing performance and multiple constraints upon agricultural strategies in the forest–savanna transition zone. It has compared production systems and household level decision-making in physically and economically distinct localities. These two levels of analysis reveal that forest clearance is a crucial element in the adaptive strategies of survival. This is in the face of sometimes very adverse local and regional pressures, clearly beyond the control of individual households. At the same time, forest depletion is widely recognised by smallholder farmers as a major factor contributing to local environmental deterioration and is expressed in declining returns for work effort and growing insecurity in food production.

Considering the relative scarcity of available forest land and recent settlement in localities where forest is still abundant, the pace of current clearance should concern both farmers and policy-makers. The problem is likely to be aggravated by further sub-division of original tracts of land, growing numbers of on-estate workers, capital accumulation by a few wealthy farmers and transfer of ownership to others who resort increasingly to cleared-forest pastures. The influx and creation of a large landless population will also result in the degradation of even marginal forested areas, as was illustrated at Tepequém. Analyses show how soils can deteriorate dramatically, even in reputedly fertile areas, once the forest cover has been removed.

In areas where natural resources have been more depleted, as at Taiano, farming conditions are increasingly risky and demanding. Economic returns have decreased to the extent that they threaten self-sufficiency in food supplies, curtailing the capital accumulation necessary to change to more suitable and profitable farming systems. Without external subsidies capital saving is now unlikely to occur. In those frontier areas where forest is still abundant all the ingredients of the same degrading processes are at work. Clearance is fuelled by pressure from adjacent savanna-based ranches in their search for dependable sources of fodder for cattle rearing, which is still perceived to be the best form of land use.

Interviews have shown that farmers are well aware of the resource deterioration that they are causing. They have developed complex forms of production to secure a livelihood, with intricate cropping patterns, multiple husbandry, the use of riverine and forest products and off-farm employment such as on ranches, in *garimpos* and urban jobs. Smallholders have tried to respond to changing conditions in an unstable climatic environment, with few government aids, a dearth of infrastructural and service facilities and severe market disadvantages. Even in these circumstances they show a willingness and ability to incorporate modern technology where it can be seen to be economically justified – such as the planted pastures at Tepequém or bush-cleared fields of Taiano. Many farmers were testing and in some cases planning to expand cash and fruit crops, despite the adverse problems of water deficiency, pests, road conditions and periodic labour shortages.

With the present infrastructure, market conditions, land zoning and titling policies, the most logical option is to clear the forest even further. Cattle husbandry seems to develop not so much as a result of government support or market demand but because the alternatives are less attractive and viable. Should current policies and economic conditions continue, the Amazonian forests, even where they are a scarce commodity and highly valued, will continue to be felled and will ultimately vanish by default.

Figure 5.5 Aerial view of Taiano, one of the older colonisation schemes in Roraima.
Source: Peter Furley.

NOTES

1 Soil samples collected by the authors were analysed and interpreted by Phillips (1990) and Furley.
2 This had an added value during the gold-rush since it could be used for mining on a *garimpo* (see Chapter 7).

ACKNOWLEDGEMENTS

Grant assistance was received from the Federal University of Pará, the Emilio Goeldi Museum, the Royal Geographical Society and ORSTOM. Grateful acknowledgement is made of logistical and other assistance provided by the Royal Geographical Society's Maracá Rain Forest Project, and by Brazilian Government agencies, notably SEMA and INPA.

6

AGRICULTURAL DIVERSIFICATION

The contribution of rice and horticultural producers

Christopher Barrow and Andrew Paterson

INTRODUCTION

The focus of attention now turns from ranching and smallholder subsistence farming, the mainstays of land economy in Roraima, to the more specialised concerns of rice production and the growing of horticultural crops. The chapter concentrates on agricultural development centring on the region around Boa Vista.

Despite the predominance of ranching in the economy, those employed in cultivation and logging accounted for nearly 50 per cent of the workforce in 1970, whereas those in livestock-related employment accounted for only around 10 per cent. Indeed, by the early 1970s, there was a shortage of trained personnel in the agricultural sector. Interviews in the Boa Vista region in the late 1980s suggest that cultivators were dependent for almost 70 per cent of their labour on family (particularly sons of 7 to 15 years old), but only 20 per cent of ranchers depended on such labour. The former were mostly smallholders, whereas the latter were mainly larger landowners. Field interviews in the late 1980s indicated this pattern still prevailed and that only about 15 per cent of those employed in cultivation, livestock-rearing and forestry were women.

In spite of having a smaller proportion of the workforce, livestock production has always been greater, in terms of size and economic value, than other agricultural activities (Table 6.1). The last agricultural census in 1980 showed that only a fraction of the farms were engaged in horticultural production (0.43 per cent), occupying an even smaller proportion of the agricultural land (0.04 per cent), although the area and importance of horticulture has been increasing over the past decade.

AGRICULTURAL PRACTICES IN THE BOA VISTA REGION

In the area approximating to the satellite scene illustrated in Plate 3.4, there are examples of forest extractivism, traditional shifting cultivation,

Table 6.1 Economic activities in the agricultural sector, 1980

Activity	Number of establishments	Percentage of total	Total area (ha)	Total area (%)
Agriculture	2,578	69.23	564,135	23.29
Livestock	902	24.22	1,761,854	72.75
Relating to cattle	73	1.96	56,406	2.33
Horticulture/floriculture	16	0.43	986	0.04
Forest extraction	85	2.28	23,930	0.99
Aviculture	70	1.88	14,529	0.60
Total	3,724	100	2,421,840	100

Source: Censo Agropecuária 1980 (IBGE 1983).

subsistence smallholdings, cattle ranching, the use of *várzea* and *cerrado* lands for arable cultivation and, more recently, an increase in the number of horticultural holdings.

Extraction of forest products

The indigenous people practise various forms of shifting cultivation of rootcrops such as manioc, whilst conserving useful wild palms (Barrow 1990; Balik 1982) and other trees which provide fruits, oils and construction materials (National Academy of Science 1975; Goodland *et al.* 1978; Myers 1984). The most important palms used by the Yanomami, for example, include the peijebaye (*Bactris gasipaes*), species of Mauritia, Orbignya, Euterpe and Astrocaryum (Anderson and Anderson 1983).

The extraction of forest products remains a small-scale but important activity both for the indigenous groups and for immigrant settlers (Sternberg 1987; Gradwohl and Greenberg 1988), especially in the Caracaraí region and in the forests to the west of Maracá Island. The most important forest products include:

Borracha de Seringa (Rubber)	*Hevea brasiliensis*
Balata (Latex)	*Manilkara bidentata*
Camaru (Cumaru)	*Coumarouna odorata*
Canarana	*Panicum spectabile*
Castanha (Brazil Nuts)	*Bertholletia excelsa*
Copaiba	*Copaifera multijuga*
Sorva (Latex)	*Couma utilis*

The collection of balata, sorva latex and Brazil nuts are the most important of these activities in Roraima. Rubber was extensively gathered during the Second World War, but when demand declined Brazil nuts became more important. The 1980 Agricultural Census estimated that some 85 collecting groups were involved in the extraction of Brazil nuts from approximately 23,000 hectares. Latex collection is undertaken during the driest months

(October to May) and Brazil nuts during the wet season (June to September); there was a small export (sorva 2.8 per cent and Brazil nuts 3.5 per cent of the total agricultural figure) in the mid-1980s.

The extractors of forest products are amongst the poorest sections of the population, frequently existing on very basic diets deficient in protein and minerals. In 1970, some 60 per cent of the population of the *município* of Caracaraí were engaged in the extraction of forest products. Almost always they are tied economically to landowners and river traders (*proprietários dos regatões*), to whom they are frequently indebted.

Traditional agricultural smallholdings

Prior to the colonisation schemes, smallholder agriculture in Roraima was dominated by subsistence farming, with manioc the principal crop followed by maize. Most of the production was undertaken by *caboclo* farmers along the riversides or to a lesser extent on *várzeas*. Today, one of the most common agricultural systems in the Boa Vista region is shifting cultivation of rain-fed rice, maize and manioc. Typically, a plot of 2 to 5 ha in size is cleared of grass and scrub between October and March using slash and burn techniques. The plot is planted with all three crops, more or less at the same time, soon after burning and just before rain. The manioc cuttings are planted about 2 m by 1 m apart with the rice and maize seeded between. Any surplus is sent for sale to the Boa Vista, Caracaraí and sometimes even the Manaus markets.

Individually farmers vary their cultivation strategy to suit local soil conditions, which can be very heterogeneous. On sandy soils, maize is planted first just before the rains. Rice is planted 15 days, and the manioc 20 to 45 days, after the maize (Roraima, EMBRAPA/EMATER 1977). On clay soils, rice is planted first, just before the rains, then maize about 15 days later and manioc 20 to 45 days after the rice. Nowadays even small farmers are likely to use improved seed varieties. Manioc is almost always processed locally, in the nearest village or on the farm, to make manioc flour. Typically, production would consist of only one or two harvests before soil exhaustion and weed growth force clearance of a new plot. Data on productivity is given in Table 6.2. In some localities, rice, maize and manioc

Table 6.2 Production of rice–maize–manioc in a shifting cultivation strategy

Crop	Actual productivity (kg ha⁻¹ yr⁻¹)	Potential (kg ha⁻¹ yr⁻¹)
Rice	1,500–1,600	1,700–1800
Maize	900	1,100
Manioc (as processed *farofa*)	3,900	3,500–4,000

Sources: Roraima, *EMBRAPA/EMATER* 1977; Roraima, *EMATER* 1981.

shifting cultivation has been modified by the introduction of a 'rotation' system which extends the period for which the cleared plot can be cropped.

Ranching

In recent years agricultural extension services have promoted a number of improvements. Until recently, there was no attempt to re-seed pastures; periodically, the *cerrado* is still burnt to promote fresh grass growth and woodland is cleared to add to grazing land, again using fire (see Chapters 4 and 5). The result is open-*cerrado* grassland (usually *Trachypogon ligularis* and *T. vestitis*), with occasional fire-resistant trees (*caimbé* or *Curatella americana*). There is little dairy production, the main goal of the low carrying capacity native grasslands being ranching for the production of beef. However, in low-lying areas, valley bottoms and ponds, water may be retained all the year round, and so watering cattle may not be such a problem (such sites may also have potential for simple pump irrigation). Vegetation in these depressions is generally dominated by the buriti palm (*Mauritia flexuosa*) and often includes the perennial semi-aquatic herb *aninga* (*Caltha lutea*). Many of the seasonal ponds are markedly round, typically from 100–200m in diameter lying at the head of drainage systems. These characteristic features of *cerrado* landscapes (*cabeçeiras*), resemble the *dambos* commonly found in parts of Africa (Zambia and Zimbabwe).

Small-scale arable farming

The development of arable crop production has been constrained by natural limitations such as poor soil, unfavourable climate and swamp or seasonally flooded terrain. In addition, the poor road system, the relatively small size and vulnerability of the consumer market, lack of capital and initial resources by immigrant settlers, difficulties of obtaining land tenure or even access to available land (and therefore opportunities to obtain to credit), periodic labour problems and, finally, the inadequacy of technical assistance, have all contributed to the difficulties facing the small-scale arable farmer.

In the early 1970s, much of the state remained undeveloped, though inhabited by a number of indigenous groups, a situation still characteristic today in the southern and western areas where dense forests and heavy rains inhibit agricultural settlement (Zimmerman 1973). Between 1950 and 1980, the number of agricultural enterprises in Roraima increased from 445 to 3,742, and the amount of land occupied from 595,795 ha to 2,463,106 ha. Zimmerman's report (1973) states that of the 1,953 mixed arable and ranching enterprises registered in 1970, only 132 properties with an area of 506,902 ha possessed definite title. Therefore, 93 per cent of the farms occupying around three-quarters of the land possessed no documentation to guarantee occupation. The lack of title deeds to property still remains one of the major barriers to development in Roraima and, indeed, in the Amazon Basin as a whole.

Contribution of colonist settlements

The agricultural colonies (*colônias*) produce a large proportion (if not all, in some cases) of the major crops marketed in Roraima. The fall in production in 1983 was due largely to exceptionally bad climatic conditions in 1982. Table 6.3 compares the estimated total production by the colonies with the total state production.

Table 6.3 Production of agricultural colonies compared with total production of major crops

| | Total prod. colônias (tonnes) | | | 1980/81 total prod. state (tonnes) | Percentage of 1983 colonial to 1980/81 state prod.* |
	1955	1964	1983		
Rice	900	1,619	11,633	44,830	25.9
Maize	63	697	8,775	14,479	60.6
Beans	22	58	694†	673†	100.0†
Manioc	–	–	58,409†	50,313†	100.0†

Source: *Censo Agropecuária* 1980 (IBGE 1983).
Notes: * Attempts to indicate proportion of the state production represented by the colonisation schemes.
† These figures are dubious; figures for the *colônias* exceed those for the state, implying that the total production comes from the *colônias*. The totals from the *colônias* have been estimated from the production of 37 schemes.

Table 6.4 Crop rotation and the percentage of farmers practising rotation

Crop	Percentage of farmers
Maize–rice	21.9
Manioc–rice	0.0
Manioc–beans	1.2
Beans–rice	2.4
Maize–beans	15.9
Maize–rice–manioc	42.8
Maize–manioc	1.2
Maize–rice	1.2
Maize–beans–manioc–rice	11.0

Source: ACAR/RR Field Research, 1973.
Notes: In the cleared rain-fed areas, near Mucajái for example, the farmers mostly use a maize–rice–manioc rotation together with water melons.

The ACAR Report (Zimmerman 1973) gives a breakdown of the major crop rotations within the state. At that time 42.8 per cent of those interviewed followed a maize–rice–manioc rotation (planted in that order), and 21.9 per cent followed a maize–rice rotation. Beans are normally grown with maize (Table 6.4).

Rice production on *várzea* lands

There is apparently little or no tradition of *várzea* cultivation in the region. Very low yields were obtained and total loss by flooding was an ever-present threat. Over the past 50 years, most land development has taken place in

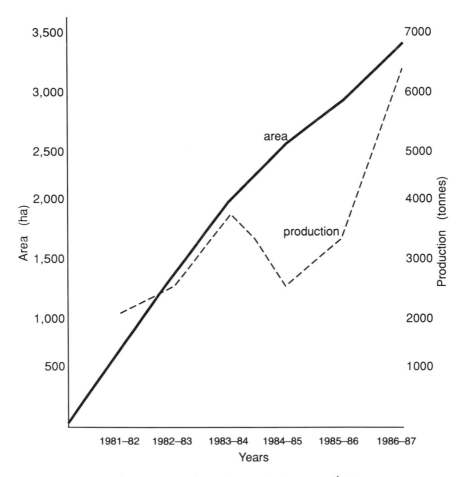

Figure 6.1 Irrigated rice areas and production in the PROVÁRZEAS Programme, Roraima, 1981–86.
Source: ASTER, Boa Vista, 1987.

areas accessible by road or track. River transport, at least above Caracaraí, is little developed and only in the last five years has there been any land development in *várzea* areas, and that only in places with road access. However, in northern Roraima studied, the development of *várzeas* in the last five years has been rapid, rising from below 500 ha in 1980 to over 3,000 ha in 1986. This relates to the high rates of *várzea* forest destruction identified in Chapter 3.

How much of the rice production in Roraima comes from *várzea* areas is difficult to establish. However, *várzea* productivity is higher than rain-fed rice productivity and there has been considerable expansion of production in these areas in recent years. Figure 6.1 indicates the rapidity of expansion in irrigated rice production and most of this is from *várzea* areas. Recent estimates suggest that there are approximately 360,000 ha of *várzeas* suitable for rice cultivation in the state (Ribeiro *et al.* 1985). The area, production and yield of the main crops, including rain-fed and irrigated rice in 1986 are shown in Table 6.5.

The main problems faced in developing *várzea* rice in the Boa Vista region (see Figure 6.2) are (a) high land clearance, levelling and fertiliser costs, (b) high labour costs, (c) rice prices prone to fluctuation, and (d) difficulties and cost of obtaining credit. In spite of these difficulties, *várzea* rice met 30 per cent of the market demand in 1986 (Santana 1986).

Currently, the main *várzea* rice production areas lie between the Rio Branco and the BR 401, along the Rio Uraricoera, Rio Murupú and Rio Cauamé; there are further promising areas along the Mucajaí and Amajari rivers (Santana 1986). Rice production (not all on *várzeas*) is important in

Table 6.5 Area, production and yield of main crops in Roraima, 1980–86

Crop	1980	1981	1982	1983	1984	1985	1986
				Years			
Rice							
Area (ha)	17,314	45,512	15,680	6,884	8,758	9,726	8,349
Production (tonnes)	25,718	44,830	18,529	7,158	15,409	15,686	7,038
Yield (kg ha^{-1})	1,485	985	1,182	1,759	1,613	843	n/a
Maize							
Area (ha)	6,024	13,473	3,751	1,877	7,366	8,665	6,254
Production (tonnes)	5,762	14,479	2,513	591	6,106	7,183	3,902
Yield (kg ha^{-1})	956	1,074	670	315	829	829	624
Beans							
Area (ha)	380	1,150	733	290	797	983	874
Production (tonnes)	162	673	302	120	391	482	261
Yield (kg ha^{-1})	442	583	412	414	490	490	299
Manioc							
Area (ha)	1,224	3,826	2,800	4,045	3,195	1,557	2,216
Production (tonnes)	17,508	50,313	38,768	56,007	44,288	21,558	16,525
Yield (kg ha^{-1})	14,303	13,150	13,846	13,846	13,846	13,846	7,456

Source: IBGE/GCEA.

Figure 6.2 The *várzea* rice scheme to the east of Boa Vista across the Rio Branco.

Figure 6.3 Ricefields at the *várzea* scheme. Photograph by ASTER staff, July 1985.

the areas around Taiano, Alto Alegre, Tucano, Tres Corações and Serra da Lua.

Since the early 1980s, there has been good progress in selecting suitable rice varieties and soil improvement techniques for local conditions. *Várzea* soils along the Rio Branco and its tributaries are quite easy to clear and grade (Figures 6.3, 6.4). The lack of large trees, at least in the predominantly savanna areas, eliminates the necessity for stump removal – a problem in parts of Pará and Amazonas (Barrow 1985). The soils, usually a mix of fine sand and silt, resist compaction, even when repeatedly ploughed or graded, and they also tend to form a crust which resists rain-splash erosion. However, the soils are less fertile than other *várzea* areas of the Brazilian Amazon. They tend to be deficient in organic matter and often have a high content of aluminium (pers. comm. ASTER; Cordeiro 1984). To improve such soils, most growers add artificial fertilisers, which are at present a costly input. A typical application would consist of 150 to 300 kg ha^{-1} of NPK 4–28–20 +Zn at sowing, and 50 to 80 kg ha^{-1} of urea in the early growing season (Cordeiro and Mascarenhas 1983; Cordeiro 1984; Ribeiro *et al*. 1985). With these levels of application, some growers have obtained 6,000 kg ha^{-1} yr^{-1} (rice varieties GA 3473, BR-1, RGA-409). The maximum yield in research station trials has been 7,963 kg ha^{-1} yr^{-1} according to Ribeiro *et al*. (1985). Growers planting seed at 120 kg ha^{-1} and using the fertiliser dressings indicated above, will generally get yields between 2,000 and 3,600 kg ha^{-1} yr^{-1}, reasonable yields compared with similar sites elsewhere in Amazonia (although along the lower Amazon and whitewater tributaries, less fertiliser application may be needed).

The strategy adopted for *várzea* rice farming resembles that used in Pará. The *várzeas* are flooded between April and late September. When exposed, between October and March, there can be extended periods with little rainfall, so irrigation is necessary. Already quite level and with a cover of grass and light scrub, the land is easily graded to form areas that can be flooded simply by pumping in water with a diesel-powered pump. Only one harvest a year is obtained at present; the rice varieties grown are IRH4GA-409 and IRGA-410, both reaching about 50 cm in height and with a 115-day growing season.

The labour requirements for irrigated *várzea* rice are between 15 and 30 man days ha^{-1} yr^{-1} (Coelho 1982). Some growers pre-germinate rice seed but most sow directly onto the plot. The fields are flooded to just over 5 cm depth about 5 days after sowing and this is maintained for about 40 to 45 days. After this time, the water level is held at 15 to 20 cm and then, 20 to 25 days before harvest, the field is completely drained (Coelho 1982). Some growers have had pest problems, mainly from *Lagarta militar* (*Spodoptera frugiperda*), *Lagarta clasmo* (*Elasmopalpus lignosellus*) and *Scaptocococores castanea* – grub and plant bugs. Capybaras and rats can also be a nuisance, and birds take some of the grain.

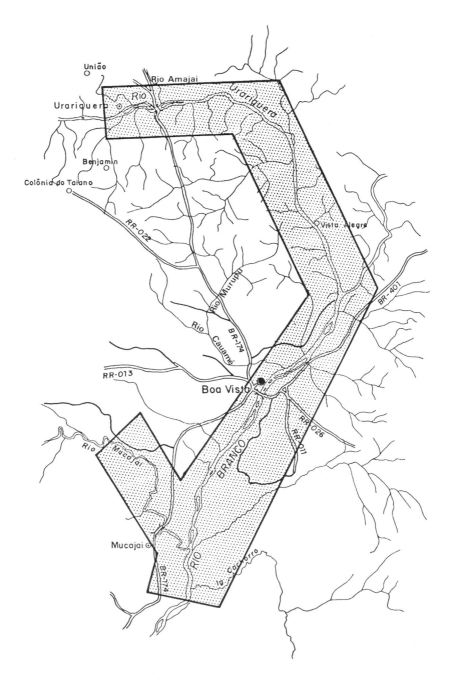

Figure 6.4 PROVÁRZEA – likely areas for rice expansion (shaded).
Source: Santana 1986.

Given access to credit and suitable land, there is no reason why small-scale farmers should not produce rice or other crops using tractors provided by co-operatives. However, it seems unlikely that larger landowners would welcome such competition for the more extensive stretches of *várzea*. Assuming that to be the case, smallholders will have to use smaller patches of *várzea* (which are less suited to mechanised cropping), or seek to establish pump-irrigation on *terra firme* using water from ponds or dammed streams. EMBRAPA in Belém have made progress with the introduction of the aquatic plant *azolla*, as a green manure for fertilising rice 'polders' (field visits 1985). If this proves to be successful, and becomes a widespread practice, it might help smallholders avoid the use of expensive chemical additives.

Other forms of rice production

Pump-irrigation on the cerrados

Trials at EMBRAPA's research station at Água Boa, close to Boa Vista, suggest there is some potential for developing irrigated rice on native grasslands (*campos*) by pumping water from ponds or earth-dammed streams. Water is applied with simple sprinkler equipment and yields are between 4,500 and 6,500 kg ha^{-1} yr^{-1} (using varieties IRGA-9 or IRGA-10; field observations, 1987). Tubewells may also be practical, as ground water is often only a few metres below the surface.

Production linked to pasture improvement

There are an estimated 36,000 km^2 of native grassland in Roraima. Since 1977, there has been increasing interest in mechanised large-scale rice production using rain-fed *sequeiro* rice. In 1979 20,000 ha of grassland was planted with *sequeiro* rice, which gave a crop of 30 million kg. By 1981, two-thirds of the state's rice was grown by mechanised large-scale growers,

Table 6.6 Rain-fed (*sequeiro*) rice varieties and yields trials (Rio Caume region)

Variety	Height (cm)	Average production (kg ha^{-1} yr^{-1})	Growing season (days)
IAC-5128	70.6	1,647	109
IAC-165	68.0	1,640	94
IAC-47	67.0	1,547	109
IAC-25	71.6	1,485	89
IAC-164	67.0	1,466	91
Amarclaō	82.0	1,400	110
Cateto	71.0	1,142	114

Source: Roraima, EMATER 1981: 8.

Table 6.7 Trials of rain-fed *(sequeiro)* rice on typical yellow *cerrado* Latosols near Boa Vista

Rice production (kg ha⁻¹ yr⁻¹)	Value Cr$ (Sept 1982)	Fertiliser cost Cr$ (Sept 1982)	Profit Cr$ (Sept 1982)
2,137	85,480	24,000	61,480
1,688	67,520	24,000	43,520
2,005	80,200	34,000	46,200
2,038	81,500	24,000	57,500

Source: Gianluppi *et al.*, 1983.

and small-scale commercial production now seems overshadowed. *Sequeiro* rice is harvested in September following a 90- to 124-day growing season and the yield is between 1,100 and 2,500 kg ha⁻¹ yr⁻¹ (varieties IAC-47 and IAC-25) (see Tables 6.6 and 6.7). The area cultivated rose by nearly 70 times from 1974 to 1981 and production increased in the same proportion – in other words an increasing quantity but virtually the same levels of productivity per hectare.

The principal problem is the inherent variability of *cerrado* soils, which are not only nutrient-poor but have subsoils high in aluminium. The latter tends to cause shallow-rooting and drought vulnerability (Furley and Ratter 1988; Furley 1990). The Dark-Red or Red–Yellow Latosols are farmed for rice whilst other soils, particularly the quartzitic sands, are totally unsuited to cultivation. Without fertilisers, *sequeiro* rice yields would be below 1,500 kg ha⁻¹ yr⁻¹ and would be uneconomic. An application of 60 kg ha⁻¹ yr⁻¹ P_2O_5 and the same quantity of K_2O boosts yields to between 2,600 and 2,700 kg ha⁻¹ yr⁻¹ which should be sufficient to sustain production (Couto and Alves 1981).

An interesting possibility is to link rain-fed *sequeiro* rice production with pasture improvement. Trials at the Água Boa research station were well advanced in the late 1980s, and economic appraisals suggest that it would be worth giving incentives for farmers to adopt this strategy; already it has been adopted by some of the larger landowners. Typically, the grassland is cleared, ploughed and seeded with rice, a forage legume (usually *Stylosanthes capitata*) which resists drought as well as fixing nitrogen, and a grass – typically *quicuio do amazônia* *(Brachiaria humidicola)*. After about 120 days, the rice harvest helps pay for the land preparation and fertiliser treatment and the improved grazing is able to carry more cattle than the unimproved *cerrado*.

A *Brachiaria/Stylosanthes* mix provides good quality forage, which in trials and in the experience of local farmers resists overgrazing and permits a greater carrying capacity. Trials with this mix near Boa Vista, which had run for about five years at the time of the field survey, showed little sign of pasture degradation in spite of heavy grazing. The disadvantage is that this type of pasture improvement will mainly benefit the larger landowner, unless the smallholders can form co-operatives or produce forage for the larger *fazendeiros*. A further possibility for smallholders to develop a sustainable system might be a crop rotation using elephant grass (*Pennesetum purpurum*) or the shrub legume *Tephrosia* with millet. This could provide a reasonable source of forage for sale.

HORTICULTURAL PRODUCTION

The present analysis of the horticultural sector is assessed in terms of the scale of production rather than precise figures (Roraima, CEPA 1983). Discrepancies almost certainly arise where no distinction is made between the figures for products grown for the market and those which are grown for home consumption.

It is useful to set the scene by comparing horticultural with other agricultural production (Table 6.8). Whereas 15 per cent of agricultural establishments are less than 2 hectares, with 41 per cent between 2 and 5 hectares, 25 per cent between 5 and 10 hectares, and 15 per cent from 10 to 50 hectares in size, horticultural enterprises are consistently smaller. Around 23 per cent are less than 1 ha in size, 41 per cent range between 2 and 5 hectares, and 23.5 per cent between 5 and 10 hectares. Only about 11 per cent of horticultural establishments, and 5.4 per cent of agricultural enterprises, belonged to co-operatives or growers' organisations in 1980. Nearly all agricultural establishments at that date used contract labour but only one unit was horticultural. On the other hand, 47 per cent of horticultural enterprises used irrigation, compared with less than 1 per cent for agricultural farms. Similarly, over 80 per cent of all the horticultural producers used chemical and organic fertilisers compared with only 3.5 per cent of agricultural enterprises. Approximately 62 per cent agricultural and 82 per cent of horticultural establishments recorded using crop protection materials. A comparison of these figures illustrates the greater intensity of husbandry and capital investment in the small horticultural sector.

Horticultural crop production had reached only a limited stage of development in the 1950s and the sole market of any significance was Boa Vista. Our survey showed that six horticultural enterprises existed at that time, of which four were managed by Japanese families.

Table 6.8 Area and number of establishments principally devoted to agricultural enterprises

	Number of establishments	Total area (ha)
livestock	3,742	2,463,107
arable + livestock	2,658	632,348
arable	2,585	72,941
horticulture/floriculture	17	7,790

Source: *Censo Agropecuário* 1980 (IBGE 1983).

Vegetable crops

Increased immigration over the past twenty-five years has created a demand for fresh produce which results in frequent shortages. This has resulted in the importation of vegetables from Venezuela and the creation of the agricultural colony of Monte Cristo (see pp. 176–7). The vague definition of horticultural crops makes it difficult to assess the actual area under vegetable production although it is known that pumpkins, tomatoes, cabbages, green peppers and lettuces are produced in significant quantities (and roughly in that order of importance).

A crude estimate of the total annual production of the major vegetable crops would be about 1,000 to 1,200 tonnes, mostly from the colony of Monte Cristo. This would indicate a low per capita consumption, a fact recognised by the extension services which are endeavouring to improve nutrition through educational programmes and home visits. Vegetable production is more intensive than for all other crops, but yields are frequently reduced through pest and disease infestation before they can be sprayed. Despite this, much of the produce is of good quality, and as Roraima's population continues to increase, particularly the urban concentration around Boa Vista, the market is continually expanding.

The principal vegetable crops are sown towards the start of the wet season in February and March and harvested where possible before the heaviest rains occur. A second sowing can be made in September or October where the soil has remained moist or where irrigation is available.

Beans, both fresh and dried, are a staple food throughout Brazil and are grown by most farmers, with varieties of *Phaseolus albus* most common. Members of the *Cucurbitaceae* are popular vegetables in the state, the most popular being the *abobora*, a small pumpkin with estimated yields of about 15 t ha^{-1}. The *pepino* (cucumber) is grown frequently on trellises, and *maxixe* (probably *Cucurbita anguria*) is a small spiny cucurbit locally popular in soups and stews, very easy to grow and gives high returns. Water melons (*Citrullus sp.*) are one of the principal cash crops of the region and a source of income for settlers in the forested areas of the south. Sweet melons are also popular and are often grown under cover during the rainy months to extend the harvesting season. The most serious problems threatening cucurbitaceous crops are mildews, which can be devastating if controls are not available.

Tomatoes are one of the most important market crops. There are two main sowing times, March–April and September–October (the beginning and end of the rains). The tomato is sensitive and subject to fungal infections, but preventative measures are not usually practised due to lack of materials. Yields of 20 t ha^{-1} are estimated by ASTER. This is low in comparison with the growers at São Bento (Rio Uraricoera) who claim yields of 50–60 t ha^{-1} harvest^{-1}. Aubergine yields are estimated to be about 30–36 t ha^{-1} harvest^{-1}. Chilli pepper is an important crop with a high market value but does not appear to be grown in any quantity although it does exceptionally well at São Bento.

Lettuce is an important and easily grown salad crop throughout the state. Normally it is raised in nursery beds and then planted out in raised beds accessible to irrigation. No serious pests and diseases were reported, but losses of 20 per cent would be considered normal. Yields of up to 26 t ha^{-1} are reported by ASTER. The climate of Roraima is not ideally suited to the production of brassica crops but despite the high temperatures, *couvé* or leaf cabbage, and a form of savoy cabbage, *repolho*, are popular crops. The coldest and most suitable time of the year is during the rains when minimum temperatures average 22–27°C during June, July and August. It is likely that seeds originated at the CNPH (Centro Nacional de Pesquisa de Hortaliças) in Brasilia. Yields are typically 11.3 t ha^{-1} leaf cabbage and 20 t ha^{-1} or more from the savoy cabbage.

Carrots, radishes and onions are widely grown at Monte Cristo. The soils of Monte Cristo are very good for root crops as they contain virtually no stones and little clay. With the addition of organic matter very high yields are possible. One farmer stated that he was getting a carrot crop of 40 t ha^{-1}, whereas ASTER estimates were 16–20 t ha^{-1}. Onions appear to be one of the most profitable and easy crops to produce. The production of *cebolinha* (spring onion) as a herb has always been important, and several thousand bunches are produced monthly for the fresh market. Although the potential yields of 10 t ha^{-1} are much lower than the national average of 40 t ha^{-1}, they are rather higher than in many developing countries with similar growing conditions. On two other farms at Monte Cristo it was evident that serious disease problems were present – probably fungal infections – and no control had been possible due to lack of materials. Sprinkler irrigation was being used and it is quite likely that this will have exacerbated the problem. Additional causes will be poor crop hygiene and infrequent crop rotations.

Okra grows well but can be severely affected by virus infections and nematodes. Nematodes appear to be a serious problem generally in Roraima, and are the vectors of many virus diseases of okra, and solanaceous and other crops. The use of strict rotations and good cultural practices would assist in their reduction and increase production. Okra is an important crop at Monte Cristo with an estimated 3,600 kg month^{-1} produced in 1983. Yields were not accurately ascertained but are likely to be in the region of 450 kg ha^{-1} of young pods or 0.5–1.0 kg per plant over a harvesting period of 4–6 weeks. It is a tolerant crop and can be grown throughout the year.

Fruit crops

Virtually all fruit production is undertaken on smallholdings and few if any large plantations exist at present. The figures provided by IBGE (1983) give

Table 6.9 Producers of major horticultural crops

Crop	Number of growers	Total area (ha)
Banana	212	48,219
Cashew nut	3	360
Orange	63	30,420
Pineapple	5	736
Tomato	9	220
Horticultural crops (vegetables, etc.)	17	7,790
Total	309*	87,745†

Source: *Censo Agropecuário* 1980 (IBGE 1983).
Notes: * Number of establishments growing horticultural crops is only 8.3% of the total agricultural sector, and only 12% of the total involved in crop production. Vegetable production represents only 0.6% of the total under crops.
† The hectarage recorded for the establishments growing particular crops is the total area of the farms and cannot be assumed to be the actual area of the crop under production. Individual crops form part of the total area under cultivation.

a reasonable indication of the total area under cultivation. Bananas are by far the most important fruit followed by oranges and pineapples (Table 6.9). Of the 309 establishments referred to in the Agricultural Census of 1980 as growing fruits and vegetables, nearly 70 per cent grow bananas and 20 per cent oranges, while only 5.5 per cent grew vegetables. More than 95 per cent of the banana growers and 84 per cent of the orange growers are tenants (*ocupantes*), but some 60 per cent of the pineapple and cashew growers are owners (*proprietários*) who could afford to take the risk of producing a higher-value crop. Despite the difficulty of establishing the average size of fruit holdings, it was estimated that some 23 per cent of banana producers were between 2–5 ha and 26 per cent were between 100–200 ha. Around 33 per cent of oranges were grown on establishments of 100–500 ha. A certain amount of indigenous fruit, such as various palm fruits and *cupuaçu* (*Theobroma grandiflora*) were seen in the markets at Boa Vista, but it was not possible to establish the quantities sold or precisely where the produce originated.

The banana is probably the most important fruit grown in Roraima. Numerous varieties exist, the most popular being *mação*. About 36 per cent of the total production comes from Caracaraí and 30 per cent from Bonfim to the east of Boa Vista. Production has increased steadily, if erratically, over the 25-year period from 1961–1986. Between 1980 and 1982 banana production increased by over 70 per cent, probably as new plantings came into fruit (in 1980, 57 per cent of the plants were estimated to be immature). There was an exceptional increase in production from 369,000 to 1.935 million bunches between 1985 and 1986. It is probable that serious rice and maize losses due to poor rains in 1982 caused growers to turn to bananas. More than 85 per cent of the 1980 harvest was produced on farms of less than 10 ha. Thirty-six per cent of the crop was sold via intermediaries, 28 per cent was sold directly, and 27 per cent was consumed on the farm. The peak harvest time for bananas is from August to November, but some are available all the year round. Bananas are normally produced by small farmers and are commonly included in a banana–maize–rice production system. The bananas are capable of recovering nutrients from deeper in the soil than the main field crops, with the result that the productivity of the land is extended and a useful supplementary income obtained from sale of the fruit in the local markets (Barrow 1985). By 1984, yields had increased to 2.4 t ha^{-1} (the 1980 Agricultural Census estimated a harvest of 1,200 bunches ha^{-1} whereas a more realistic figure for smallholder cultivation would be about 625 bunches ha^{-1}).

The second most important fruit crop in the state is the orange. The 1980 Census estimated that 63 farms produced oranges, representing 20 per cent of all farms producing horticultural crops. The production for that year was estimated to be 6.1 million fruits harvested from 56 ha. These figures differ greatly from the national statistics which indicate a production of 2.6 million fruits from a slightly larger area. Production showed a substantial increase over the period 1961 to 1986. The area under production at present is hard to gauge, but IBGE figures indicate that some 905 ha were under cultivation in 1986, of which 303 ha were harvested. The same figures suggest average yields of 4.2 t ha^{-1} – which is very low but probably realistic. The 1980 Census also indicated that 25 per cent of oranges were grown on farms of less than 20 ha and a further 33 per cent on farms of 100–500 ha. Apart from a new plantation at the Fazenda São Gabriel outside Boa Vista, all the trees seen in the field were unnamed varieties, possibly grafted but more likely seedlings. Nevertheless the quality of many of the plants growing along the Mucajaí and near Maracá Island was good. However, given good quality rootstocks and improved varieties, and despite the seriousness of diseases such as Phytphthora and Gummosi, the prospects for upgrading citrus production seem encouraging. The bulk of the orange crop is harvested in September and January. Tangerines and lemons form the bulk of the other citrus fruits in the region. The 1980 production of lemons totalled over a million fruits with 304,000 tangerines. The area under lemons increased over 2.5 times between 1980 and 1984 and the production of fruit increased in the same proportion. Yields are estimated at 13 t ha^{-1}.

Production of pineapples increased markedly between 1961 and 1986, with exceptional harvests during 1984–85. This may have resulted from plantings following the bad years of 1981–83. Roraima, ASTER (1982) estimated a yield per ha of 19,460 fruits (excluding losses of 25–30 per cent). This would give a total area of 100–110 ha under production, with 60 per cent of the producers being *proprietários*. Almost certainly many more growers produce pineapples, although market supplies may only come from a limited number. The most important producing areas are at Alto Alegre (40 per cent) and Caracaraí (25 per cent). Gross margins calculated from

ASTER suggest that pineapples may not be very profitable compared with other fruits. They are, however, a useful crop and tolerant of pests and diseases in forest areas.

Very little statistical information is available on mango production, for which conditions in the state are favourable. The 1980 Census gives a production figure of 765,000 fruits which would approximate to 200–250 tonnes. All are produced from areas of less than 1 ha. The Census also shows that there are four harvesting peaks, two minor ones in February and June and two more important in October and December. Anthracnose is a serious problem, particularly when the trees come into bearing in the rainy season. There are good prospects for mango production in Roraima given the introduction of appropriate varieties, although a market for higher quality fruit will have to be found outside the state.

The papaya is an important fruit throughout the region, but few if any plantations are likely to grow high-yielding named varieties. Papayas are very easily produced and almost all smallholders have a number of plants from which surplus fruits are sold on local markets. As a source of nutrition for the poor, they are probably extremely important. The production in Roraima is estimated to be around 230,000 fruits (1980 Census). ASTER suggests a production of 6,000 fruits under good conditions, but 50 per cent of this would be more reasonable. The total area producing for the market might be in the order of 75–100 ha based on these yields. There is little doubt that yields could be readily improved with the introduction and distribution of new varieties, combined with good extension advice concerning propagation and maintenance of promising lines. There has been considerable success with Hawaian varieties grown in the Tomé–Açu area of Pará for sale in São Paulo.

Almost no information is available on the extent of avocado production despite its popularity. It is unlikely that there are any named varieties in use at present, but attempts to increase production were observed in the nursery at Monte Cristo. Production in 1982 was approximately 17–25 tonnes, very small quantities which signify little attempt at formal production. The harvest could be increased significantly, but once again it would need the development of a market either inside or outside the state.

Distribution and marketing of horticultural produce

Distribution and marketing in Roraima remain at a basic level for the major crops, fruits and vegetables. Most the state's agricultural population maintains itself with subsistence cultivation, which does not lend itself to the production of marketable surpluses at a commercial scale. Livestock and rice production are economically the two most important agricultural activities.

Horticultural crops tend to be of high value compared with the main arable crops. The demand for such produce has increased greatly with the large number of better-off immigrants (such as civil servants) into the *município* of Boa Vista. On the whole, local supply remains insufficient to cater for this demand despite the production from the colonies of Monte Cristo and São Bento. A substantial amount of produce is still imported from the south and Venezuela. Increasing the supplies of perishable produce from the south of the state, will be difficult until the highway from Manaus is completely paved.

The majority of growers have little concept of modern post-harvest handling and packing techniques. Tomatoes are delivered in open wicker baskets containing 20–30 kg of fruit, much of which gets damaged by compression. Most other vegetables, such as peppers, carrots and onions, are delivered in sacks or large polythene bags. Fruits such as water melons, bananas and oranges are sold loose on the floor of the market. Some of the smaller and higher value fruits are sold in bags from the stalls. Vegetables and dried goods are sold from raised display areas and quality varies according to the time of day or week. Deliveries might only be made once or twice a week so the consumer has to visit the market regularly to get the best-quality produce.

The cost of living in Roraima is generally high when compared with the remainder of Brazil. Road access from the main centres in the south is difficult and costly, so that perishable goods are inevitably flown in. It is likely, therefore, that the low consumption figures of fruit and vegetables are influenced by the high consumer prices as well as by traditional diets.

An examination of the available producer and consumer market prices confirms their volatility, particularly between 1985 and 1987 when there was a major currency crisis in the country. Some consumer prices flattened out (at a high level) due to price regulations enforced by the government, including, for example, oranges, tomatoes and rice. This did not last long however. In some periods, producer market prices exceeded those in the consumer market. Production throughout the state is very seasonal, reflecting the prevailing climate. Combined with the geographical isolation of Roraima, this results in wide price fluctuations in both consumer and producer markets.

Transportation of produce is a major problem for horticultural producers. The larger growers possess their own vehicles and, in the case of Monte Cristo for example, do help those lacking transport. More distant colonies such as Taiano are dependent upon private carriers. Middlemen are naturally in evidence, but their precise methods of operation and the margins on which they operate could not be established. Many of the smallest growers and extractors of forest produce are totally dependent on them, especially where produce has to be transported by river.

It was hoped to illustrate the relationship between prices and the severe climatic conditions which occurred during 1982–83, when many of the major crops such as maize, rice and beans were destroyed by exceptionally heavy rains. In fact, substantial variations were not evident from recorded market prices, which were probably kept more or less under control by imports.

Costs of production in Roraima are influenced by isolation, which means that virtually all bulky materials such as fertilisers and agro-chemicals have to be brought in by road. Transport charges are high and this is reflected in the retail prices. For example, the largest egg producer in Boa Vista has to import most feedstuff from São Paulo. Such a producer is fortunate in having a captive market for what would otherwise be a non-commercial operation. Not only are production costs high, but actual availability may be limited for any number of reasons – a frequent one being the impass-ability of the road from Manaus. This lack of availability of essential inputs was graphically illustrated during 1987, when potentially good crops of onions and tomatoes were being destroyed by disease due to shortages of pesticides and fungicides.

In theory, a wide range of materials is available in Boa Vista, but in fact only a few are readily available and many are unsuitable for the horticultural crop producer. Even if they can be purchased, the costs are likely to be prohibitive for the small-scale farmer. The problem is not one of ignorance; growers on the colonies around Boa Vista are well aware of what is required and make every effort to obtain the necessary materials.

Table 6.10 Crop gross margins (per ha in Cz$)

	Production costs	Cz$ gross returns	Gross margin
Lettuce	67,207	427,680	360,716
Aubergine	56,283	180,000	123,716
Onion fol.	67,025	1,126,400	1,059,375
Onion	118,764	160,000	41,236
Carrot	117,501	180,640	63,139
Coriander	71,695	910,080	838,385
Cabbage fol.	43,973	227,700	183,726
Cucumber	66,348	180,000	113,652
Capsicum	82,670	780,450	697,780
Chillies	59,380	200,000	140,620
Savoy cabbage	77,659	390,225	312,556
Tomato	107,800	201,600	93,800
Rice (irrigated)	9,418	12,500	3,082
Maize seed	2,663	4,200	1,536
Maize fresh	11,298	16,000	4,702
Pineapple	58,252	72,196	13,944
Orange	10,924	26,000	15,076
Papaya	45,267	?30,000	?15,267
Water melon	34,818	?32,920	?1,898
Passion fruit	38,253	120,000	81,747
Banana	No figures available for production		

Sources: Costs of production: ASTER 1982; returns: *Boletim Agropecuário* 1986. Average producer market prices for the state.
Note: Costs of production relate to post-1986 (March) when the *cruzado* was introduced.

Gross margins for the major fruit and vegetable crops have been calculated on the basis of production costs provided by ASTER, and the average producer-market prices provided mainly by the monthly *Boletim Agropecuário* (Table 6.10). An attempt has been made to estimate the most profitable crops, based on these figures. The detailed costs and returns are not important but the estimated margins give some idea of the most profitable crops in the mid-1980s. The best ones, per hectare, such as onions, are normally grown on very small areas and have a limited market. Increased production can rapidly glut the small consumer demand. There is, therefore, a predictable gap between supply and demand, but supplies are sufficient periodically and for some of the crops.

TECHNICAL ASSISTANCE, EXTENSION SERVICES AND RESEARCH

Until 1972 there was no government extension service. In that year ACAR/ RR was established and became ASTER/RR in 1976 (Associação de Assistência Técnica e Extensão Rural Roraima) and was affiliated to the national organisation EMATER. ASTER exists to plan, co-ordinate, and execute technical assistance programmes as well as the transfer and dissemination of technology to improve the way of life of the state's agricultural producers; it now (1988) has 27 offices throughout Roraima.

ASTER has been involved in a number of programmes:

1 the development of the *várzea* lands for crop production (PROVÁRZEA);
2 incentives for rubber planting (PROBOR);
3 incentives for fish breeding (PESCART);
4 livestock developments involving dairy, meat production, and animal traction;
5 technical assistance to medium and small producers of maize, rice, and beans;
6 development of the horticultural sector, particularly food supplies to urban areas;
7 assistance with the improvement of nutrition levels and general health among the rural producers (BEM-ESTAR SOCIAL);
8 rural youth programmes;
9 rural credit programmes;
10 attempts to develop a 'collective conscience' among farmers to assist with the reduction of production costs.

Like many organisations in Brazil, the extension agency, ASTER, is short of recurrent funds and often unable to maintain a project once started. The staff of most agencies appear to be keen, knowledgeable and aware of the problems which may be caused by the implementation of government policies.

173

At least five experimental stations have been established in the state, none of which is exclusively devoted to horticultural crops, though the one located near the colony of Monte Cristo tends to specialise in vegetables. It was not possible to ascertain accurately the present state of their development but CEPA (Roraima 1984) provides the following information:

1 *Monte Cristo*. Area: 460 ha; 12 km from Boa Vista. On dark-yellow latosols of medium texture and low fertility. Typical of *cerrado* conditions. The work here has been mainly on rice and livestock, but the potential for soya, sorghum and cowpea has also been investigated.

2 *Água Boa*. Area: 1,200 ha; 30 km from Boa Vista. Sandy soils of low fertility, again typical of the *cerrado*. Work on livestock and pasture improvement has been undertaken here as well as extensive trials with forage legumes, exotic forage crops, and with pump irrigation for arable crops (for which there may be potential in seasonally flooded *cerrados*). The station is run by EMBRAPA.

3 *Confiança*. Area: 300 ha; 100 km from Boa Vista. A forest area on red–yellow latosols in the *município* of Bonfim. Work has involved the development of technologies for the small farmer using minimum inputs. Crops examined include rice, beans, cassava, guaraná (*Paullinia sorbilis*), together with some pasture research.

4 *Bom Intento*. Area: 1800 ha; 35 km from Boa Vista. This station contains both *várzea* and *cerrado* land and is part of the old *fazenda* of Bom Intento. Work on livestock has been undertaken together with crops such as irrigated rice, rubber, fruits, vegetables, and other annual crops.

5 *Serra da Prata*. Area: 600 ha; 70 km from Boa Vista. This station lies in the transitional area between the *cerrado* and forest in the *município* of Mucajaí. Dairy production is an important part of the work, together with rubber and guaraná.

The field stations have co-operated in a number of national programmes for the improvement of the major crops: rice, beans, maize, cassava, rubber; livestock development; diversification programmes; *cerrado* development; humid tropics production systems, and fruit production.

As part of a national programme for the development of horticultural research it was proposed to examine the behaviour of carrot cultivars in Roraima when planted at different seasons; to effect the introduction of lettuce in Roraima and to appraise different cultivars; to introduce and appraise cabbage cultivars; and to investigate the viability of tomato planting in Roraima.

It was not possible to establish how successful this programme has been, but there was certainly evidence at the colony of Monte Cristo of successful on-farm vegetable trials being undertaken with the assistance of ASTER. This work is discussed in the section on vegetable production (see pp. 166–7).

The role of EMBRAPA

A large new headquarters with well-equipped laboratories has been built for EMBRAPA on the Caracaraí road outside Boa Vista. There are close links with the network of EMBRAPA establishments throughout Brazil, and especially with CPAC in the Federal District (devoted to agro-pastoral developments in the *cerrado*), and with the vegetable research and production station CHNP in Brasilia. Discussions with staff at these research institutes suggested that no programmes are yet geared specifically towards conditions in Roraima.

CASE STUDIES OF AGRICULTURAL COLONIES

Comparisons may be made between crop production under very different conditions at many *colônias* near Boa Vista (a *cerrado* area) and Caracaraí (forested area). These have been settled mostly by poor families from the north and north-east of Brazil who have traditionally practised subsistence agriculture, and who have little familiarity with modern production techniques.

The problems of the forested Caracaraí area are extreme when compared with the *cerrado*, in that maintenance of cultivable land is made difficult by the high rainfall which encourages rapid re-growth of secondary vegetation and annual weeds. This has been exacerbated by present government policy (albeit unwittingly) of mechanically clearing the entire vegetation cover. The colonists are then presented with several hectares of cleared forest with little means of cultivation. The soil is frequently eroded or degraded before any protective crop cover can be established. Development of sound market-based production in these areas will be difficult due to the distances from the main markets and lack of reliable transport. Most of the settlers maintain themselves on a subsistence basis and sell surpluses when these are available. Water melons, for example, are one of the most useful market crops in the Caracaraí area. It is in these areas that forms of sustainable agriculture need to be encouraged. Where possible, plots should not be fully cleared and, initially, only sufficient should be prepared for the family, allowing the remainder to be cleared gradually. A policy of leaving the most valuable trees, or ultimately replacing them with those of greatest economic value, is recommended.

Those settled in the *cerrado* regions near Mucajaí are relatively well-situated from a marketing point of view, but require considerable management skills to cultivate the soils successfully. Others have started fish farming and some have small herds of dairy cattle. Livestock production may in fact represent the best way of utilising the available land rather than crop production, which is limited by the seasonal shortage of rainfall (see Chapter 5). The basic problems experienced by the settlers in both the forested and *cerrado* areas are more or less the same. The most suitable area

175

for horticultural crop production seems to be in the region near the Venezuelan border at Pacaraima which is at a significantly higher altitude. ASTER has apparently assisted with production trials of beet, carrots, peppers, and potatoes in that area, but no information on performance and yields was available. An additional benefit is that the area is connected to Boa Vista by the main road from Santa Elena in Venezuela.

The colony of Monte Cristo

The only colony that devoted itself to the production of horticultural crops was established in 1976 when an area of 460 ha was allocated for the purpose close to Boa Vista. This became the colony of Monte Cristo and was initially divided into plots of 10 hectares for the production of vegetables for the capital, although pig and poultry production was also envisaged. Some of the more successful growers have, however, increased the size of their lots beyond the initial 10 ha by absorbing those of settlers who have left the colony. The area is typical of *cerrado* conditions with acidic, dark-yellow latosols of medium texture and low fertility and high aluminium content (Roraima, CEPA 1984).

Vegetable production is largely seasonal with less being produced during the period of heaviest rains (April to August). A reservoir in the centre of the colony provides most of the water supplies and many of the holdings have tubewells to supplement their supplies. Water availability appears to be reasonably constant. Although water is available, a serious limiting factor for many growers is the lack of irrigation equipment. Many growers are unable to cultivate the full extent of their plots for this reason, and the production areas therefore tend to be near the dam or tubewell. Due to lack of irrigation, horticultural crop production in the *cerrado* areas is largely determined by rainfall patterns, with a rain maximum in June (300–350 mm) and minimum in January (c.30 mm). Access roads are generally good even though they are unsurfaced and produce can be marketed during most of the rainy season. Six Monte Cristo growers were visited at the end of March 1987. Three of these were small family-run holdings of about 10 ha and three were larger operations of 20–25 ha. The latter employed Guyanese labour in addition to their own families. Due to the poverty of the soil, almost all vegetables are grown in raised beds which facilitates the incorporation of additional organic material. Seeds are usually sown in nursery beds and then transplanted.

A number of growers have erected substantial wooden structures clad with plastic in order to extend the cropping period into the rainy season. Melons and tomatoes are grown in this way. Others have erected structures to shade crops during the hottest periods – carrots and lettuce for example. The more substantial growers frequently have an additional source of income such as poultry, and have often installed extensive irrigation systems which enable greater utilisation of their holdings. The utilisation of fertilisers and crop protection materials is limited due to problems of availability in Boa Vista, and crop losses are consequently high.

Most of the farmers in the colony grow some fruit for home consumption, but there were no orchards of any size. Various citrus, unnamed seedling mangoes, graviola, 'ata' (custard apple), passion fruit (*maracujá*), melons and water melons are the most popular. Expansion of fruit production is unlikely until sufficient water is available throughout the year – citrus plants on one holding were suffering severely from both lack of water and mineral deficiencies induced by the acid soils.

One of the causes of low yields and poor quality produce has been the lack of high quality seeds and plants. Most seeds are home-saved, but the extension services have been assisting farmers on the colony by conducting on-farm trials using improved Brazilian varieties – in particular of carrots and onions. A limited range of fruit trees is also available at the colony. In April 1987 the only plants available for distribution appeared to be lemon, grafted onto seedling rootstocks, and banana (Missouri variety which is resistant to Panama Disease). Avocado seedlings were being grown under shade, as were some young grape vines. The quality of the plants and the maintenance of the nursery was at a low level.

The colony of São Bento (Passarão)

This is a relatively new colony established in 1982 at the settlement of São Bento beside the Rio Uraricoera some 50 km from Boa Vista. While it is within the *cerrado* region climatically, it benefits not only from the presence of the river but also from the better alluvial soils on which it is located. For these reasons it is a much more suitable location for the cultivation of horticultural crops than Monte Cristo.

The colony is within relatively easy reach of Boa Vista by a good unpaved road. There are some 10 families of settlers in the colony each having a minimum of 20 ha. Water is readily available from the river but it has to be pumped some distance, especially when the level is low; lack of pumping and irrigation equipment has restricted the development of some plots. The pH of the soil is low (pH 4–5) but the growers feel that it needs very little fertiliser apart from some additional phosphorus to produce good crops. The organic matter levels are far higher than at Monte Cristo and the soils are easily worked.

As at Monte Cristo, the settlers are experienced growers originating from various parts of Brazil. They are as innovative as they can be under the present circumstances and one farmer has started exporting green peppers to Manaus and Brasília – the first such enterprise seen during fieldwork. The most popular crops are green peppers, chillies, tomatoes, maxixe, beans, and okra. No one grew carrots or onions at this colony. Fruits appear to do well and there were some exceptionally large mango trees at the original settlement, and passion fruit also grows well. New plantations of até, graviola, guava, and citrus were evident and there were tamarinds which fetch a good price in Boa Vista.

The main crops were:

1 Maize. Three varieties – 'Regional Crioulo' (120 days growing season), 1,200 kg ha^{-1} yield; 'BR 5105' (70 days), 3,200 kg ha^{-1}; and 'Regional' (90 days), 1,500 kg ha^{-1}.
2 Green peppers. The crop takes some 60 days to first fruiting and yields 2–5 kg per plant over a period of 4–5 months.
3 Chillies. This crop does exceptionally well here using a good local variety. It is a three-year crop selling in Boa Vista.
4 Tomatoes. Tomatoes are a 90-day crop using a local round fruited variety and yields of up to 60 t ha^{-1} are claimed (about 3 kg per plant). The crops seen were grown on stakes which, although labour-intensive, provide high yields of good quality fruit. They are also growing an introduced variety which is considered to be very disease-resistant.
5 Passion-fruit. This is one of the most successful and profitable fruit crops grown in the area. About 300 plants are grown per hectare yielding some 120,000 fruits after three years. Yields here are possibly a little below the estimated average of 180,000 fruits per hectare (ASTER).

In general no serious problems were seen apart from one tomato crop which had been destroyed by a fungal infection akin to powdery mildew (*Leveillula taurica*). Manure is available, and due to good soil conditions the farmers do not need to apply large quantities of fertilisers.

Substantial new plantings of fruits were in evidence. The introduction of good named varieties would certainly benefit the colony, and there is little doubt that overcoming problems of credit, seeds, plant stock and other inputs, would permit considerable development.

The colony of Vila Iracema

The colony is located near a pre-existing village, about 98 km south of Boa Vista on the BR 174. There were around 1,000 colonists in 1987, settled on plots along forest feeder roads leading off the main highway towards the Rio Branco for 15 to 20 km. The plots were much smaller than those closer to Boa Vista, averaging only 2.5–3.0 ha. They were initially cleared by hand and the trees gradually burned, but they are now being cleared and windrowed mechanically by a government agency. The soils are poor and sandy, low in organic matter apart from the first few centimetres, with low phosphorus levels and a sandy subsoil 15–30 cm below the surface. Most of the settlers are very poor and come from a subsistence-farming background. There has been no organised agricultural development in the area, but

ASTER have recently established an office. Many of the settlers appear to live in the village and travel daily out to farms.

The principal crops are still the traditional subsistence varieties. Typically, the cultivation strategy is to plant manioc followed by rice, maize and frequently water melons (which fetch a good market price) and other vegetables. Some sugar-cane is also grown. Many of the vegetables grow poorly, owing to the high humidity which promotes pest and disease problems. However, most of the popular fruits, such as mango, pineapple, citrus, banana and passion fruit do well, and there is scope for development. Mango and cashew are subject to attacks of Anthracnose and do better in areas with a drier climate.

It would seem that the majority of the colonists exist at a subsistence level, with little short- or medium-term prospect of improvement given current rates of financial and technical support.

ISSUES CONSTRAINING AGRICULTURAL AND HORTICULTURAL DEVELOPMENT

The major issues affecting horticulture are essentially the same as those affecting the whole of the agricultural sector. Some influence horticultural production more acutely because fruit and vegetable crops are generally grown by smallholders.

Communications

Probably the most serious obstacle to development is the lack of an efficient road system. Access to and within Roraima is difficult owing to numerous physical barriers and the long distances. Almost all heavy freight comes up from Manaus some 800 km to the south. The BR 174 is mostly unsurfaced and subject to serious flooding during the rainy season. It takes two weeks to reach São Paulo by road from Boa Vista. The Perimetral Norte has opened up limited access as far as Macapá to the west and the Rio Jatapú in the east, but the latter is more seasonal and an unreliable route for the transport of agricultural produce. International access to Boa Vista from Venezuela is possible by a good road and there is improved communication with Guyana. Numerous plans for road development exist, but the costs are likely to restrict development for a long time to come. It is obvious that encouragement is needed for the growth of non-perishable cash crops and processing facilities.

River transport is important but heavy loads can only come up river as far as Caracaraí and are then carried by road to Boa Vista. The multitude of smaller rivers up-country are vital access routes to the interior, but many are shallow, particularly in the dry season, and cannot handle large loads. Air communications to Manaus are still good, despite cutbacks over the last few years.

Crop storage

In some seasons, drying is a serious problem and mould and decay are common. There is a need for cheap drying facilities (possibly solar), similar to the type developed by EMBRAPA in Pará. Insect and rodent damage is considerable (probably over 25 per cent of harvested products), and could be alleviated by constructing storage bins and buildings. Even partial success in counteracting pest damage could represent yield increases of 10–15 per cent – a boost difficult to match by improvements in other farming techniques.

Security of tenure

It was commonly said of nineteenth-century Ireland that tenant farmers spent three years of their nine-year leases undoing the damage of the previous land user, three years farming efficiently and three years exploiting and ruining the land – in case the lease was not renewed. This illustrates the point that, without security, smallholders are unlikely to invest much time or effort or money in improving cultivation. Many of the settlers are *posseiros* (squatters) with no legal title.

Credit for small-scale and medium-scale producers

Smallholders are likely to be slow in re-paying loans, and are frequently defaulters. In drier *cerrado* areas, failure of rains, heavy wet season storms and numerous other environmental hazards may result in lost harvests. Credit provision therefore needs to be flexible if there is a serious intention to support agricultural colonisation.

Consumer market

The market for horticultural produce is more or less confined to the urban area of Boa Vista. To expand horticultural production, the market inside and outside Roraima needs to be developed. A detailed examination of potential markets, both within Brazil and in adjacent countries, might identify possible 'windows' for trading, since Roraima is the only region of Brazil north of the Equator which can grow both European-type vegetables and tropical fruits in any quantity.

Agricultural settlements and colonies

Management policies and strategies today suffer from the same problems as existed at the start of agricultural settlement over forty years ago. Many of the large 100 ha plots typical of the early colonisation schemes remain relatively backward because of a lack of capital, labour and essential inputs. Settlers have led a precarious existence, as has been detailed in numerous studies throughout the Amazon, and abandonment of smallholdings is common. The absence of farmers' organisations and co-operative groups has meant that there has been no back-up support to prevent farming failure. Land titles are uncertain and this has prevented access to any available credit schemes. Feelings of resentment against authorities has resulted in disturbances and a legacy of antipathy and distrust which prevails today. The problems have been highlighted by CEPA (Roraima 1984) together with suggested solutions, most of which directly address the constraints considered above.

Agricultural research and dissemination of new technology

Despite the presence of the national research agency EMBRAPA, only a limited amount seems to have been accomplished in the horticultural sector. The dissemination of information on methods and technology is carried out by the extension service (ASTER and equivalents). In the late 1980s, this was an efficient organisation with enthusiastic staff who were doing a good job despite extreme shortages of funds. ASTER has to rely on research developed elsewhere, particularly on methodologies developed for different parts of the country which are not always appropriate for the soils and climate of Roraima. They are additionally handicapped by the extremely low level of technical knowledge of many of the settlers and by the poor road system which hampers regular farm visits.

DISCUSSION AND CONCLUSIONS

The fact that Roraima is so isolated from the remainder of the country and the principal centres of population has contributed to the slow pace of agricultural development. The distances, acute physical difficulties, and costs of transporting heavy goods such as fertilisers and equipment have resulted in negligible progress in most aspects of crop production. A further major constraint has been the unavailability of capital inputs. Combined with poor soils (particularly for horticultural crops), these limitations have led to a situation where imports from southern Brazil and, to a lesser extent, Venezuela are required.

Cattle ranching together with rain-fed and *várzea* rice production are the main large-scale agricultural activities in the state. Overall, the agricultural systems remain at a traditional and unsophisticated level and little effective research work has been undertaken which is specifically geared to the problems of the region. The government extension service does its best to spread appropriate technical advice to the agricultural community, but is hampered by lack of funds, insufficient trained personnel and the economic and physical constraints considered above.

Crop production was not given any serious attention until the 1950s, when the first expansion of agricultural colonies took place. Some improvement in the principal food crops occurred (maize, manioc and rice), but increases in output are a result of expansion in the cultivated area rather than betterment of husbandry or levels of productivity. Boa Vista is the largest consumer of anything other than staple crops, basic vegetables and fruits. The most successful colony devoted to horticultural production is Monte Cristo. Other agricultural colonies appear to be badly affected by the poor road system and acute marketing problems.

The prospects of a significant improvement in agricultural production in the state cannot be viewed optimistically at present, due to the existing financial situation in the country and the environmental limitations of the region outlined in the first chapter. Many of the smallholders and colonists in both forest and *cerrado* areas are disheartened by the lack of assistance from the government and are leaving the land for the centres of population in the hope of better employment. Many have become *garimpeiros*, either temporarily or permanently. Paradoxically, this resulted in increased food demand and afforded the growers in the Boa Vista area a considerably expanded market for a period of several years during the gold-rush (see Chapter 7).

Discussions with the smallholders and colonists in both forest and *cerrado* suggest that many of those in the *cerrado* areas are from the southern parts of the country, although the majority come from the north-east or from Rondônia. Most rural immigrants have little knowledge of improved technology or agriculture and have very few personal resources. However, many of the growers near Boa Vista exhibit considerable ingenuity in both vegetable and fruit production despite a lack of appropriate technical assistance and planting material. Disregarding economic and other constraints, the localities visited showed potential for the expansion of fruit production (mango, citrus, banana, cashew and pineapple) as well as vegetables in general. Before this can happen, however, markets for such produce outside the state will have to be established.

One development which could take place is the marketing of potentially profitable indigenous plants which could provide a source of income for the poorest settlers and forest dwellers. This is an aspect which deserves to be followed up with further research.

In Roraima it is clear that production of rice in the *várzeas* and *cerrado*, and horticultural practices within reach of Boa Vista, are expanding. The problem is to ensure that any surpluses from traditional smallholdings or from commercial production, destined for sale in Roraima or further afield, can be sustained with a minimum of land and river degradation. There are numerous strategies suitable for the region, but all are likely to demand inputs of organic matter, agrochemicals and trained management to be successful. Unlike some *várzeas* elsewhere in Amazonia, those in Roraima receive little useful nutrient addition during floods because the river sediments are relatively infertile.

To sustain production without land damage or stream and groundwater pollution, careful monitoring will be necessary. It is likely that the use of fertilisers will require tight controls and that organic compost production should be promoted along with safer pesticide and pest control methods.

There has been very little tradition of fruit and vegetable production in Roraima and horticulture represents a minute fraction of the agricultural output. The majority of smallholders practise traditional agriculture with maize, rice, beans and manioc, a few green vegetables and some fruit trees. A decade ago, fruit and vegetable growers made up little over 10 per cent of the total engaged in crop production, with a little under 1,000 ha devoted to vegetables and over 4,000 ha to fruit, mostly banana, citrus and pineapple. As might be expected, the size of the farming enterprises was small, with 81 per cent between 1 ha and 20 ha. These proportions are unlikely to have changed much over the past decade, although the area under horticultural production has risen as a consequence of population growth and increased demand.

Horticultural growers in the areas close to Boa Vista have the benefit of a relatively large and sophisticated market, made up of immigrants from many parts of Brazil with a taste for their home produce. On the other hand, growers in the forested areas are constrained by distance from markets, poor roads and lack of transport. They remain largely subsistence farmers, with water melons, bananas and forest extracts their principal marketable products. Most fruit and vegetable crops suffer from poor post-harvest handling as well as a lack of crop protection measures. The unreliable and rough transport, the inadequate packing, and poor storing facilities lead to considerable losses. The improvement of horticultural production is in the hands of ASTER and EMBRAPA. Both are curtailed by a lack of funding and the problem of attracting and retaining high-calibre personnel.

Comparisons with other Brazilian Amazon areas

Despite its location within the Amazon Basin, Roraima is climatically distinct, as evidenced in previous chapters. The pronounced dry season in the north-east permits different cropping regimes and the growth of warm temperate plants. The forested areas south of Boa Vista to Manaus have a potential for a different range of arable and horticultural crops. Research at INPA, devoted to more typical Amazonian conditions, is concentrating upon crops more amenable to the region. These include members of the *Amaranthaceae* (*Amaranthus cruentus*, *A. hybridus* and *Celosia argentea*), members of the *Bassellaceae* (mainly for use as green vegetables), and various root crops such as some of the *Araceae* (*Philodendron spp.*, *Colocasia spp.* and *Xanthosoma spp.*). Other crops of interest include legumes such as *feijao-macuço* (*Pachyrrhizus tuberosus*) and *feijão de asa* (*Psophocarpus tetragonolobus*). Solanaceous fruits are also promising. Although many of these crops do well under humid tropical conditions, nematodes, mites and fungi (*Pythium* and *Sclerotium* especially) can cause heavy losses.

Prospects for development

Expansion and improvement of horticulture and agriculture generally depends upon the elimination of constraints – which are frequently logistical. Natural limitations such as poor soil cannot be improved without road access for agro-chemical inputs. With better and flexible credit opportunities, market enlargement and increased availability of technical assistance and materials, significant improvements could be made.

Many of the smallholders and colonists are capable of producing high-quality crops despite lack of inputs and losses during and after growth. The ability to protect crops is affected by poor transport; this applies particularly to those farmers at some distance from Boa Vista. Furthermore, much of the production remains at subsistence level owing to a lack of reliable markets. Both ASTER and EMBRAPA are doing their best to improve farming activities, but they are hampered by a lack of funds for staff and research. The potential for expansion and crop improvement can be readily seen, but until the major constraints are addressed it would not be appropriate to make specific recommendations.

ACKNOWLEDGEMENTS

The authors gratefully acknowledge the assistance given to them by many individuals, bodies and agencies in Roraima, particularly ASTER and EMATER; by the Royal Geographical Society, SEMA/IBAMA, INPA and the British Academy.

7

LAND-USE PRESSURES AND RESOURCE EXPLOITATION IN THE 1990s

Gordon MacMillan and Peter Furley

INTRODUCTION

The past fifteen years have brought unprecedented social and economic change to Roraima. Planned government directives, promoting agriculture and ranching alongside newly constructed highways, provided the initial catalyst for this transition. Latterly, however, land development has been of a more spontaneous nature, as migrants have flooded into the state especially during the 1987–90 gold-rush on the lands of the Yanomami Indians. The lure of mineral wealth has attracted an even greater number of immigrants than during the earlier search for land, and is principally responsible for Roraima's spectacular population growth over the past decade – higher than for any other Brazilian state. The rapidly expanding *garimpos* (mines) absorbed capital and labour from the ranching and agricultural sectors and accentuated a pre-existing trend towards urbanisation. The impact of these changes has affected both urban (Abers and Lourenço 1992) and rural economies (MacMillan 1993a and b), but has also had political repercussions that remain an important influence on land development in the state.

The reasons for the unique course of events in Roraima are examined in this chapter, with an assessment of their impact on current and foreseeable developments. A number of significant factors may be distinguished, including the political complexion of the state versus federal government, the impact of transport growth, the changing but continually vulnerable fortunes of the agricultural and ranching sectors, the critical role of mineral exploitation and the startling urban expansion, notably in the capital Boa Vista. Such an approach illustrates how the environment and the different forms of land examined in this volume interact in their regional context, and how they may influence the evolution and development planning of the state.

POLITICAL FORTUNES AND MISFORTUNES

Roraima's transition into an independent state in 1990 led to the first state elections, which saw an immediate change in the power base. Traditionally, political power had been wielded by the wealthy *fazendeiros*. The governor had been appointed by the federal government and, like other senior members of the state administration, was typically a member of either the local land-owning class, or the armed forces, or both. Roraima's newly conferred status changed this picture. Apart from reinstating a previous governor, Brigadier Ottomar de Souza Pinto, to his former position, the elections revealed a significant erosion in the power of the traditionally influential local elite. The established faction of *cerrado* ranchers had been losing their political hegemony to professionals and entrepreneurs in Boa Vista since the mid-1970s. The rapid expansion of the mining and urban sectors during the gold-rush further marginalised the political influence of the old elite, even though a few of its members were able to maintain authority by diversifying their business interests into the growing sectors of the economy. A similar set of processes operated at Itaituba in Pará during the early 1980s (Miller 1985).

Although Roraima's transition to a federal state signifies a greater degree of political autonomy than in its earlier territorial phase, it is bound to be economically uncomfortable as the federal government progressively reduces its funding. Self-sufficiency in financial terms is supposed to be achieved by 1996, according to Freitas (1991) and the Programma Especial de Investimentos (Roraima, SEPLAN 1991). Whilst the fledgling state seeks new possibilities to expand the local economy (and thereby acquire economic autonomy), the federal government maintains a controlling influence over the development of the greater proportion of natural resources. This is because indigenous areas, military training grounds, nature reserves, lands within 100 km of federal highways, as well as a 150-kilometre swath along international frontiers, are all administered by the federal government. A report commissioned by the Ottomar administration concluded that 62.8 per cent of Roraima remains under the control of the central government in Brasília (Roraima, Governo de Roraima 1991). This prompted the state to solicit (so far unsuccessfully) for the transfer of land from the control of the central government, notably along the BR 174 and BR 210 (Figure 1.14). Of particular significance is the state government's very limited control over the mineral sector, since all productive *garimpos* (with the unique exception of Tepequém) are situated on existing or proposed indigenous reserves. The central government can therefore regulate the exploitation of mineral resources in Roraima by effective policing and monitoring of the indigenous reserves.

These domestic political changes are occurring against a background of international interest in land development in Roraima. The gold-rush attracted world attention and provided a focus for international media concern over the environmental and social impacts of resource exploitation in Amazonia. According to the Environmental Defense Fund (EDF), one of the United States' most influential NGOs, 'no other single issue is more indicative of the [Brazilian] government's political will to defend the Amazon and its inhabitants than the fate of the Yanomami territory' (EDF 1991). Although the decision to stop the gold-rush and demarcate the Yanomami reserve was not directly incorporated within debt negotiations (as might characterise a 'debt-for-nature' swop), Brazil's dependence upon foreign credit makes her particularly sensitive to external political pressure. Thus, even though media attention on Roraima may wane in the aftermath of the United Nations Conference on Environment and Development (UNCED), held in Rio de Janeiro in June 1992, it is likely that the concerns expressed by foreign governments over Amazonian development will continue to influence the course of events in Roraima. Understandably, this exogenous pressure has fuelled an ongoing discourse on Amazonian sovereignty, which is vigorously defended by regional politicians. In Roraima, these arguments are voiced by a union of ranchers and *garimpeiros* who have formed the 'movement against the internationalisation of Amazonia'. They have been especially outspoken against the Catholic Church and other foreign-backed non-governmental organisations which are working in defence of Indian rights.

Land-use changes in Amazonia are particularly sensitive to changes in the political climate because government policy is the driving force behind the occupation of the region. The road construction programme, land colonisation schemes, the Calha Norte project[1] (Pacheco 1990; Albert 1992; Allen 1992) and rural credit schemes, are all examples of government directives determining land utilisation in Roraima. Frontier occupation in the interests of national security has been a cardinal feature of Brazilian policy in the Amazon and is central to the political development of Roraima. Its roots lie in the 1937 Constitution of the Vargas administration, which set up federal territories largely for this reason. Although planners intended that economic growth could be pursued by means of a rapidly assembled infrastructure of government organisations, the regional development agency SPVEA, and its successor SUDAM, allocated very little of the total available funding to Roraima. This amounted to only 0.65 per cent by 1985 (Freitas 1991). The area was marginalised economically and politically and it could be argued that, up to around 1975, it served as little more than a frontier outpost for the armed forces.

The centralisation of authority prevented residents from participating in local government until the passing of laws in 1969 (Decree Law 411). This decree, which reasserted the goal of effective occupation of the federal territories, notably in 'open spaces and frontier zones', created government secretariats and established municipalities, leading to the participation of local people in the political system for the first time. The gradual decentralisation of power gathered pace during the democratisation process known as the *abertura* (the 'opening') initiated in 1979 during the Figueiredo presidency.

In spite of these moves towards greater freedom and political autonomy, the armed forces continued to exert considerable influence over land development in the state: of the 24 state governors from 1943 to 1985, 14 held positions in the armed services. These militarists are the main protagonists of the *desenvolvimentista* (developmentalist) ideology, which was the policy initiated by Vargas and can be identified in all of subsequent development programmes which have affected Roraima. The strategic Development Plan (1968–70), the three National Development Plans (1971–74,1974–79,1979–85) and the Calha Norte Project (1985–90) all embody notions of frontier occupation.

Having identified the political influences on land development in Roraima, it is appropriate to consider the current pattern of land use in the state and to evaluate the possible modifications to this picture over the next decade. Changes in the nature and structure of transport lie at the heart of land-use change, as transport costs determine the economic viability of many productive enterprises.

TRANSPORT GROWTH

Whereas land links have been extended and improved continuously over the past two decades, movements by air have witnessed fluctuating fortunes. The isolation of Roraima from the rest of Brazil, even the remainder of the Brazilian Amazon, has led to an unusual dependence on air traffic. The journey to Manaus takes something like two hours by air but 36 hours without delays on the road system. However, air transport is currently experiencing cutbacks after the affluent gold-rush years, which created a surge in demand for both air-taxis and domestic flights. The number of daily services to Manaus has fallen from four to three, air-taxis now receive little demand and the weekly flight to Caracas was cut in 1989. The international status of Boa Vista was finally terminated by the withdrawal of the Guyana connection in February 1992. Nevertheless, the road link has now been completed between Georgetown and Boa Vista, constructed by the Brazilian group Paranapanema at the invitation of the Guyanese government. Consequently, communication between the two countries is actually increasing. Similarly, planned improvements to the BR 174 between Boa Vista and the Venezuelan border at Pacaraima could augment cross-border trade and, despite the appalling state of the road at times, can be envisaged as a powerful force in opening up the north of Brazil to the south of Venezuela.

Upkeep of and investment in federal highways is the responsibility of central government and, in Roraima, much of the maintenance is carried out by the army (notably by the Sixth Frontier Battalion). In 1991, an extensive programme of road improvements was launched, and by the end of the year all of the federal highways had been considerably upgraded. This greatly reduced the time taken on long journeys, but heavy seasonal rainfall will ensure that the improvements are only temporary unless the road surfaces are either paved, or regularly regraded. A 70-kilometre stretch of the BR 174 Manaus–Venezuela highway between Boa Vista and Mucajaí is at present the only paved segment of federal highway in Roraima. There is political pressure from both Roraima and the neighbouring Venezuelan state of Bolivar, to metal the road completely. However, while a surfaced road runs from Caracas to the boundary with Roraima at Santa Elena, the Brazilian government has not yet fulfilled an agreement it signed to continue the paved section as far as Manaus.[2] In spite of an agreement by the Brazilian government to undertake the project, it has yet to be fulfilled. It is estimated that one paved lane in either direction will cost US$72.6 million and it is therefore unlikely to be completed during the current years of recession (Roraima, SEPLAN 1991). Early in 1992, SUDAM agreed to fund an extension of the BR 174 southwards from Boa Vista by 100 km so that it reaches the inland port of Caracaraí. In the absence of foreign investment, it is possible that the full extent of the BR 174 will be paved gradually over a number of years in this fashion.

On the assumption that this and other road construction projects form a central pillar in Roraima's economic development, an asphalt-producing plant has been installed in Boa Vista's industrial zone (details of the limited commercial activities within this zone are given in Anjos Filho 1991). In January 1992, it was reported that the first output has been targeted for the 98-kilometre-long Boa Vista-Alto Alegre road (RR 205), which by the end of February 1992 had been totally regraded and was awaiting its new surfacing. Despite this, it is quite likely that the asphalting programme will not go ahead due to lack of funds. Pressure to pave the 194-kilometre-long Boa Vista-Bonfim road (BR 401) has increased too, now that the Guyanese road between Lethem and Georgetown is passable. Additional demands have been created by the rapidly growing urban system in and around Boa Vista.

Since the opportunities to exploit mineral reserves are constrained by federal ownership of land, the state government has had to look to agro-pastoral expansion as a possible economic driving force. Consequently, despite its problematic history and disappointing results in the Amazon, land colonisation is being encouraged and the extension of feeder roads represents a state priority. Some of these are being developed to serve areas of land which have been occupied by squatters. A good example is the southerly extension of the principal feeder road of the colonisation project Confiança III (part of the INCRA rapid settlement project known as PAD Barauana/Juaperi), so that it connects with the BR 174/210 at Novo Paraiso (Figure 1.5). In fact this is the original planned course of the BR 174 and the completion of the link will reduce the road distance between Boa Vista and Manaus by an estimated 50 kilometres, as well as bypassing the time-consuming ferry crossing of the Rio Branco at Caracaraí. Not only would this alter the pre-existing transport network considerably but it will inevitably open up a new area of agriculture and ranching, typically accompanied by timber extraction. Indeed, INCRA has already planned a colonisation project with the aim of settling 2,000 families along this extended stretch of road between 1993 and 1998. The initial maps of this imaginatively named PAD Nova Esperança, indicate that some of the feeder roads will border the reserve of the Wai-Wai indigenous group (Figure 2.1).

It is unlikely that Roraima will experience the same levels of deforestation and rural violence that have accompanied road construction in Mato Grosso, Rondônia, Pará and, more recently, Acre. For while the opening up of logging tracks and feeder roads are well recognised precursors for land invasion and forest clearance (Moran 1981, 1990; Smith 1982; Leite and Furley 1982; Barbira-Scazzochio 1980; Léna 1988; forthcoming), the intensity and speed of such developments relate to the political and economic context in which they occur. Roraima's geographical isolation from the principal markets of Brazil, coupled with her comparatively poor timber resources (by Amazonian standards), did much to prevent the development of an explosively speculative land market during the first phase of road-building in the 1970s. At the start of the 1980s, the tax incentives for ranching, initiated through SUDAM, were withdrawn. This, together with a nationwide recession, has reduced the availability of venture capital for speculative investments in Amazonia. Moreover, the ubiquity of easily accessible public lands (*terras devolutas*), will ensure that even if land disputes do increase they can be rapidly defused (as happens at the present time, on the whole), and may not necessarily result in violence. This does not mean to imply that contemporary developments in Roraima will not exacerbate existing social problems and increase environmental degradation, it is simply to suggest that they are unlikely to be as marked as elsewhere in Amazonia. Clearly, government planning can do much to minimise these potential impacts: two priorities are the demarcation of the remaining proposed Indian reserves (notably those of the Makuxi[3]), and a change in the legislation that currently recognises cleared forest as a claim to land holding.

AGRICULTURE AND THE IMPACT OF *GARIMPAGEM*

The examples quoted above illustrate the familiar model that expansion of the agricultural frontier is both a product of, and a motive for, the development of new roads. Throughout the 1980s, the settlement of migrant families has slipped to a large extent beyond the control of government organisations as colonisation has acquired a momentum of its own. Newly arriving families take the initiative of extending feeder roads in order to stake claim to unowned public lands situated on the periphery of existing colonisation projects. Both Alto Alegre and Apiaú have expanded considerably (see Figure 1.14) in this fashion, and a new trail of settlement is currently occupying the area between the Mucajaí and Apiaú rivers. The *posseiros* (land squatters) are lobbying the state government to convert their forest track into a feeder road, thereby both linking the two colonisation projects and opening up a considerable area of hitherto uncultivated lands on the eastern boundary of the Yanomami indigenous area.

The current state government is particularly sensitive to the requests of the colonists, as it was Brigadier Ottomar de Souza Pinto who oversaw the establishment of many of the colonisation projects during his first term in office (1979–82), and their votes provide much of his electoral support. Appealing to the colonist vote in his 1990 election campaign, he declared his intention to distribute land to a further 50,000 families should he be re-elected.[4] During 1991, the Office of the Secretary for Agriculture settled over 260 families in pre-existing colonisation schemes and planned to colonise a further 4,000 in 1992 in the two projects of Apiaú and Trairão (pers. comm., Secretaria de Agricultura, Boa Vista, January 1991). The Trairão project is located at Tepequém, one of the settlements examined in some detail in Chapter 5. Up to the present, the majority of those settled are families that were already resident in Roraima (many of whom entered the state during the gold-rush), but the government is bracing itself for a new wave of immigration. The unit handling the registration and welfare of migrants has been revived and is currently dealing with the first arrivals from the north-east of Brazil.

Although government support for land colonisation throughout the Brazilian Amazon is consistently inadequate, Roraima's smallholders are exceptional in receiving help for transport and marketing of their produce. As is shown in Chapter 5, these are key factors in the fortunes of colonist households. The state government has responded to the demands for help by providing a truck service that visits colonisation projects in the area around Boa Vista, offering free transport for agricultural produce to the weekly market in the capital. Although this project was conceived in 1987, it has received inconsistent support from the state government, being abandoned during the gold-rush years. However, it has now been resuscitated following the re-election of Ottomar to governor in the wake of the gold-rush.

Alternative public sector investments in smallholder agriculture, such as rural credit schemes, extension services, and research, are regarded by many of the colonists as being of little value. This is mainly because the financial and bureaucratic procedures of the institutions that provide these services make them inaccessible to those with least resources. Consequently, the 1980s have seen little change in agricultural technology and the crops planted by smallholders, despite the provision of government-funded research (EMBRAPA) and extension services. As both a cause and effect of this, the most significant advances in increasing the state's agricultural output, such as the developments in irrigated rice, and horticultural production described in Chapter 6, require levels of capital investment that are beyond the financial limits of the majority of colonist farmers. Rural credit does little to make the new technology more accessible or encourage crop diversity, as proven land ownership (possession of *titulo definitivo*) remains a prerequisite for any loan. In 1991, only 15 per cent of contracts arranged under the FNO (Fundo Constitucional de Financiamento do Norte) credit scheme (which is specifically designed for small and medium farmers), were applied to properties smaller than 500 hectares (Banco de Amazônia, pers. comm., January 1992).

As an insurance against the hazards of smallholder farming, and the inconsistencies of government agricultural policy, a large percentage of colonists have adopted alternative sources of 'off-farm income'. The *garimpo* and the expanding urban economy have been the most attractive options throughout the past decade (Abers and Lourenço 1992). In order to keep these options open there is a reluctance for colonists to commit themselves to longer-term farming strategies, such as the planting of perennials, that might help in improving plant diversity and tie them more closely to the land. As a result of this management practice, short-term economic, climatic, or political anomalies have an exaggerated impact on agricultural production as people fluctuate between the agricultural and non-agricultural sectors of the economy. A good example of this is the 1987–90 gold-rush, during which an estimated 49 per cent of colonist households sent at least one of their members to the *garimpos*. As a result of this exodus of agricultural labour, the production of crops such as dry rice, maize and bananas declined (Table 7.1).

Table 7.1 Areas planted under annual crops compared with population in Roraima, 1985–90

	Area planted under annual crops (ha) 1985–90					
	1985	1986	1987	1988	1989	1990
Crop type						
Banana	500	2,419	2,504	3,005	1,859	1,989
Dry rice	9,124	8,238	5,546	5,728	3,855	3,000
Irrigated rice	602	1,200	1,341	1,490	2,775	3,025
Maize	8,665	6,044	6,753	6,952	3,807	3,318
Manioc	1,537	2,583	1,177	1,567	1,974	2,132
Total	22,448	20,484	17,321	18,752	14,271	13,439
1985 = 100	100	100	85	92	70	66
Population	159,600	163,571	177,693	233,422	275,082	276,322
1985 = 100	100	103	111	146	172	173
Planted Area/Pop.	0.128	0.125	0.097	0.080	0.052	0.048
1985 = 100	100	98	76	62.5	41	37

Source: Secretaria de Agricultura and SUCAM, Boa Vista.

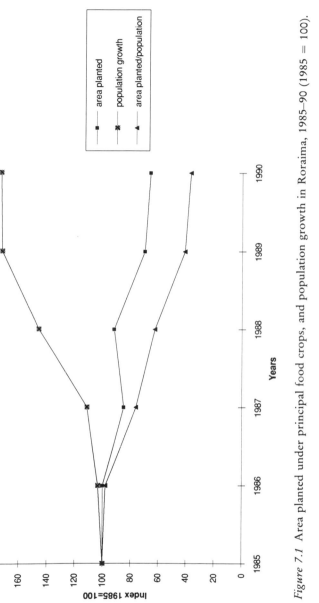

Figure 7.1 Area planted under principal food crops, and population growth in Roraima, 1985–90 (1985 = 100).

Between 1985 and 1990 there was a 35 per cent decrease in the area planted under annual crops (data from the Secretary of Agriculture, Boa Vista) while the population increased by 73 per cent (SUCAM data) – see also, Figure 7.1. This reduced the index of hectares planted per individual of population by 63 per cent over the same period (1985 = 100 per cent for all data). Equally dramatic was the return to the land at the end of 1990, stimulated by both the closure of the *garimpos*, and promises made by the recently elected Ottomar administration to improve feeder roads. The result was a pronounced increase in the area under cultivation, and a bumper harvest which, in the absence of storage facilities, was allowed to flood onto the contracting Boa Vista market.

Notwithstanding its environmental repercussions (see Fearnside 1989), numerous problems still hamper smallholder agriculture in Roraima as has been indicated in earlier chapters of this book: the inappropriate transfer of technology, the concentration on a handful of annual crops, the difficulties of obtaining credit, transport costs, and a shortage of warehousing space with drying and moisture control facilities are among the most prominent. It is therefore, a matter of concern that the state government is committed to an expansion of agricultural colonisation in Roraima, without addressing these fundamental issues.

RANCHING AND LIVESTOCK

As demonstrated in Chapter 2, cattle ranching on the natural grasslands of the *cerrado* has been the linchpin of Roraima's economy for over 150 years, and beef was one of the state's principal exports up until the end of the 1970s. However, throughout the 1980s changes in both the absolute size and distribution of the beef herd have altered this picture considerably. Although there is no exact record of the number of cattle in the state, SUDAM estimates a fall from 360,000 head in 1980 to 220,000 in 1991 (Brasil, SUDAM/OEA/PROVAM 1991; see Figure 2.2), representing a 38 per cent decline over the decade. Alternative data provided by the state government is more conservative, reflecting a tendency by local politicians to over-emphasise the importance of the beef sector in the state's economy. Even so, they do recognise that Roraima's herd declined by 6.9 per cent between 1980 and 1985, and fell by the accelerated rate of 8.2 per cent between 1985 and 1988.[5] The salient point is that a gradual decrease in the size of the beef herd accelerated during the latter half of the decade.

A variety of factors are responsible for this dramatic decline, some of which are related to the structure of land tenure on the *cerrado*. Traditionally, cattle were branded and left to roam across the grasslands, allowing them access to distant water courses during the dry season. With the increasing tendency to fence ranches, migration has been restricted, and on some *fazendas* (notably the smaller ones) water scarcity has become a limiting factor on beef production. Ranchers also complain that the demarcation of lands for the Makuxi indigenous group is reducing the area of pasture available to their herds. In fact, while current reserves account for less than 5 per cent of the total *cerrado* area, it is likely that potential demarcations, which could effect 50 per cent of these natural grasslands, are discouraging some *fazendeiros* from making further investments. However, this may not have such a large impact on beef production as the *fazendeiros* claim, for the Makuxi themselves are now rearing cattle through two projects financed by the Catholic Church and FUNAI (the government Indian agency).

Although the ranching community tends to blame supply-side factors for the decline in the size of the beef herd, employees in the agricultural extension service emphasise failings in the short-term management practices for the changing fortunes of the livestock market. A high percentage of ranchers have alternative sources of income (notably government jobs), and many have investments in other sectors of the economy – particularly in retail, mining, and construction enterprises. Indeed, for many ranchers, the *fazenda* is simply a mechanism for banking financial resources from their other business interests. Just as small farmers transfer their labour when agriculture is in decline, ranchers will diversify capital to different sectors of the economy as investment opportunities arise. A survey of 137 *fazendeiros* who own land on the *cerrado* revealed that 70 (51 per cent) have other business interests,[6] a feature noted throughout Central and South America. In response to the high demand for beef associated with the growing urban market in the late 1980s, many ranchers sold off considerable parts of their herds, and typically invested the proceeds in either the *garimpo* or the urban economy. This management practice was responsible for a decline in overall beef productivity, as some ranchers did not simply destroy the unproductive bulls or spent cows but went as far as slaughtering productive heifers in an attempt to raise capital rapidly.

The decline in beef production over the traditional *cerrado* areas is partly offset by expansion of the ranching frontier, spreading into the forested areas to the south and west of Boa Vista. These ranchers differ from the traditional land holding elite that dominates ranching, and represent a new migrant class of businessmen and professionals whose economic base rests on commerce, gold-mining, and, in some cases, timber extraction. Typically these ranches, which line the BR 174 and BR 210, are from 500 to 2,000 hectares in size and rear mixed-race cattle on artificial pastures. A number of colonist farmers have also acquired a few cattle (in many cases the direct result of a lucky strike in the *garimpos*), and so contribute to the increasingly significant forest ranching frontier.

The higher levels of rainfall, and the short-term increases in available soil nutrients following deforestation, permit higher stocking rates in the forested areas, at least initially, compared with those on the *cerrado*. In response to this a number of ranchers have adopted a technique of breeding livestock on the relatively disease-free grasslands, before transferring the young adults to fatten up on the artificial pastures of the forest. Cows raised in such a fashion acquire sufficient weight for slaughter in only 2.5 years compared to 4.5–5.5 in the *cerrado*. Economic advantages such as these, and modest returns from property speculation, suggest that ranching in Roraima's forested areas will continue to expand (agreeing with the conclusions of Mougeot and Léna, Chapter 5), in spite of its detrimental impact upon the region's soils and vegetation (as shown by Eden and McGregor, Chapter 4).

The new practice of fattening livestock on artificial pastures represents one of the few developments in Roraima's ranching management for decades. Even though the beef-ranching sector accounted for over 75 per cent of rural credit applied through the FNO (data from the Banca da Amazônia 1992), there is still minimal investment in pasture improvement technology, in better genetic stock and selective breeding, and in artificial insemination – any of which could improve the dismal productivity of the state herd. Not only are credits directed mainly towards infrastructural improvements like the construction of fences and corrals, but according to the local overseer of the FNO credit scheme, a significant proportion of the money spent is actually channelled by ranchers into their other business ventures and is consequently never applied as intended. The ranch, in this event, represents a means of securing access to cheap government loans which can be invested elsewhere.

Even though cattle raising in Roraima is not underpinned by the booming, speculative, land markets that have been observed in earlier decades in Pará, Mato Grosso, Rondônia, and more recently in Acre and Tocantins, it is apparent that ranch management strategies are defined largely by external economic factors. Unless this situation changes, and ranch management more closely reflects the direct economics of production, then it is hard to envisage it becoming a more efficient land use in terms of productivity. Furthermore, government attempts to restore Roraima's status as a beef exporter will remain ineffective if credit schemes permit the continued rechannelling of cheap loans into other sectors of the economy.

THE MINERAL SECTOR

The mineral sector in Roraima has been expanding progressively since the first discovery of diamonds on the Rio Maú in 1912. It is important to measure recent events against this historical perspective and recall that only a fraction of the state's diverse mineral potential (which includes gold, diamonds, cassiterite, and radioactive deposits; see Chapter 1), has so far been exploited.[7] This is because the virtual halting of the 1987–90 gold-rush was a result of political pressures, and does not reflect a substantial depletion of resources. Consequently, the current contraction of mining operations shows many signs of being little more than a short-term downturn in a generally upward-moving curve.

In fact it is only in the old alluvial diamond mines of Tepequém and the Rios Maú, Quinó, and Cotingo, that reduced output is ascribed to gradual exhaustion. This is understandable given that these mines have been producing steadily for over 50 years, the last 15 of which have seen the application of semi-mechanised machinery. Nevertheless, these traditional *garimpos* have attracted a new influx of miners following the closure of gold and cassiterite operations on the Yanomami indigenous reserve, and it is unlikely that their significant influence on the local economy will wane rapidly. This is because virtually all of these older *garimpos* are accessible by road, and a workforce exists that is prepared to work for very little pay (much of it indigenous labour from the Makuxi communities). Consequently, overheads for diamond production are minimal in comparison to those of gold-mining, which occurs in remote forested areas often only accessible by light aircraft. This relative ease of access goes some way to explaining why the Makuxi have been unsuccessful in their attempts to expel the *garimpeiros* from the proposed (but currently undemarcated) Raposa/Serra do Sol reserve.

By contrast, government intervention has greatly reduced the extraction of gold and cassiterite, which employed an estimated 25,000 to 40,000 miners in the area of the Yanomami Indigenous Reserve. It is important to distinguish between the production of the two minerals even though they are often exploited together. For although they are extracted by the same workforce using the same technology, their future development is shaped by different processes.

The mining of alluvial gold-bearing deposits was ultimately responsible for opening up the Yanomami reserve, which yielded approximately 36 metric tonnes of gold between 1988 and 1990 according to the Amazonian *garimpeiros* union (Feijão and Pinto 1990). The high value/weight ratio of the mineral, and facility with which it can be traded, generated numerous small gold-mining operations that were rarely dominated by larger mineral groups. The same factors ensure that the policing of gold-mining operations on the Yanomami reserve will remain problematic. The police and miltary surveillance is likely to impede rather than totally eradicate illegal mining. Assuming that no roads are constructed in the area in view of its now demarcated status, the number of *garimpeiros* working there is likely to fluctuate according to the intensity of police vigilance, the price of gold, the cost and availability of fuel, and variations in other sectors of the region's economy. According to the *garimpeiros* themselves, there are still sufficient gold deposits to satisfy at least another five years *garimpagem* using existing technology, and they point out that large areas remain unexplored.

Cassiterite mining is more readily concentrated in the hands of larger corporations because it has a limited number of buyers and a lower value/

weight ratio which naturally incurs higher transport costs. The extraction of high-quality cassiterite from an estimated 15,000 metric tonne deposit (Fernandes and Portela 1991) in the north-western region of the Serra de Surucucus within the Yanomami indigenous reserve, attracted substantial capital investment. By the end of 1989, two principal operations, that used 1,200-metre airstrips serviced by DC3 transport planes, came to dominate *garimpo* cassiterite production. It appears that these considerable investments received financial backing from multinational companies. However, both operations had only succeeded in expanding output to full levels of production a few months before mining activities were closed down by the federal police in May 1990.

It is worth noting that the Association of Tin Producing Countries (ATPC) was urging Brazil to restrict informal sector cassiterite production at this time. Although Brazil is only an observer at the ATPC meetings, it is the world's largest producer of tin ore and so shares the cartel's interest in maintaining high prices. Falling demand from industrialised nations and increased supply from producer countries have driven down the price of one tonne of cassiterite traded in Brazil by 63 per cent in six years (US\$10,265 in 1984 to US\$3,728 in 1990; Brasil, DNPM 1990). Indeed the prospect of increased production from the Roraima *garimpos* was considered directly responsible for a downturn in world tin prices during the second half of 1989.[8]

Overall, the contemporary economic and political climate opposes the continued development of mineral resources on the Yanomami reserve, but one must question how long this position will prevail. It may be significant that in November 1991 the government mining production department (DNPM) sent a team to survey the geology of the Surucucus region – shortly after the *garimpeiros* had been removed from the area. The subsoil wealth under the Yanomami Indigenous area can be regarded as a strategic mineral reserve which could be exploited if future economic or geopolitical forces dictate. If and when these resources are mined remains open to speculation. It is worth recalling that mining on indigenous lands is acceptable within the Brazilian Constitution if authorised by the National Congress (Paragraph 3 of Article 231), on the condition that it has the consent of affected indigenous communities. In the meantime, both *garimpeiros* and formal sector mining companies will continue to pressurise state and federal government for access to the deposits.

URBAN GROWTH AND THE DEVELOPMENT OF THE TERTIARY SECTOR

One of the striking features about Roraima is that it supports a considerable urban population and yet has no manufacturing industry. The urban population is overwhelmingly concentrated in Boa Vista, which has nearly

120,000 out of a total state population of some 216,000 (IBGE 1992). Within the *município* of Boa Vista (Figure 1.15), the urban proportion is 83 per cent. The Caracaraí *município* is 58 per cent urban, but the remaining *municípios* have low, predominantly rural, populations (Table 7.2).

There are no specific plans to encourage manufacturing, for planners are seeking to stimulate the tertiary, not secondary sector of the urban economy. This is to be fulfilled by the creation of two 'Free Trade Zones' at Pacaraima (Venezuelan border), and at Bonfim (where the road to Georgetown crosses the Guyanese frontier), which aim to follow the Manaus model of tax-free imports of selected goods for retailing. Both of the Roraima projects have been approved by Congress and were signed by the President in November 1991, but by the end of January 1992 the decrees that would establish the Free Trade Zones had still not been published. While realising these plans may attract some investment in the assembly of consumer durables, it is likely that they will do more to stimulate trade flows of retailed products between Manaus and neighbouring countries. In either case, it will expand the urban economy in Boa Vista as well as in these frontier towns.

Urban demand for electricity is currently supplied by diesel-burning generators. With existing equipment, Boa Vista can produce 80 megawatts of electricity, a capacity that considerably outstrips her current demand of 30 megawatts (pers. comm., Eletronorte, January 1992). In spite of this potential, two hydroelectric projects are planned so that future growth of Boa Vista and other urban centres is not limited by electricity shortages.

Table 7.2 Urban–rural population distribution in Roraima's municipalities, 1991

Municipalities	Total	% urban	Municipalities	Total	% urban
Boa Vista	142,813		*Alto Alegre*	11,195	
Urban	118,926	83.3	Urban	3,347	29.9
Rural	23,887		Rural	7,848	
Mucajái	13,148		*Normandia*	11,170	
Urban	5,166	39.3	Urban	1,139	10.2
Rural	7,982		Rural	10,031	
Caracaraí	8,910		*Bonfim*	9,453	
Urban	5,136	57.6	Urban	1,224	12.9
Rural	3,774		Rural	8,229	
São Luiz	9,099				
Urban	2,242	24.6			
Rural	6,857				
São João da Baliza	10,002		*Roraima*	215,790	
Urban	2,292	22.9	Urban	139,472	64.6
Rural	7,710		Rural	76,318	

Source: After IBGE 1992.

Survey work has already begun at a site on the Rio Jatapú, where four turbines of 2.5 megawatts are to be installed. This scheme is significantly larger than the originally planned 3-megawatt plant, which was designed to meet the energy needs of the southern half of the state (currently estimated to be less than 1 megawatt). The project costs are estimated at US$22 million, and the state government was looking to the *Fundo de Participação dos Estados* (FPE), a type of regional development fund, for the necessary financial resources (data from the Compania Energia de Roraima (CER), Boa Vista, January 1992).

A 150-megawatt hydroelectric plant is also planned for the Rio Cotingo, which is currently under consideration by the World Bank. Although it would inundate a relatively small area (16 square kilometres) of *cerrado* vegetation, its potential social impact is polemical due to its location on the proposed Makuxi indigenous area of Raposa/Serra do Sol. This alone makes it difficult for the state government to secure financing for the scheme, but their chances are further jeopardised because they are currently repaying a US$15 million loan to external creditors for a previous hydroelectric project at Paredão that was never completed.[9]

However, it is not only the energy sector that may be transformed by the creation of the Free Trade Zones. Agriculture and ranching could move towards a more intensive use of the nutrient-poor *cerrado* soils if currently expensive inputs such as fertilisers, pesticides, irrigation equipment, fencing materials, lime, and farm machinery, fall significantly in price. This is unlikely unless a greater demand reduces unit costs, but cheap energy provision and local manufacture or mixing plants for fertilisers could achieve a significant price reduction. Favourable yields of soya beans and dry rice have already been attained through the application of energy-intensive 'green revolution' farming techniques, but the future profitability and viability of such highly capitalised land uses rests on economic and political factors. Certainly the prospects of reduced input costs and improved transport links to the Caribbean, are encouraging some farmers to look towards export-led agricultural growth however optimistic that might appear at the present time.

LAND-USE ZONING AND DEVELOPMENT PLANNING

In an attempt to plan these rapidly changing land uses, the government is seeking more precise information concerning the region's natural resources. Dargie and Furley (Chapter 3) outline some of the many potential applications of remote sensing technology in Roraima, which is acquiring a growing importance in the land-use planning of the state. SLAR, Landsat (MSS and TM) and AVHRR data has already been used by federal government organisations to detect illegal mining operations in the Yanomami Indigenous area (FUNAI), to monitor deforestation (IBAMA/INPE), and to map promising mineral deposits (DNPM/CPRM). Of current interest to the state government is a natural resource and socio-economic survey of the Rio Branco Basin being undertaken by SUDAM within the wider study programme of the Amazonian valleys (PROVAM).

The Rio Branco survey is a forerunner to the more extensive Ecological–Economic Zoning of Roraima (ZEERR), which is currently in its initial stages. The project, costing US$3 million, is financed by the International Bank of Reconstruction and Development and is being co-ordinated by IBAMA in conjunction with the state government. It foresees the establishment of a small Geographical Information System (GIS) with image processing facilities located in Boa Vista, and aims to produce an 'environmental profile' of Roraima at scales of 1:2.5 million and 1:1 million. For the priority areas identified by the state government advisers (see Figure 7.2), mapping will be at scales from 1:1 million to 1:250,000. For areas identified as critical, it is intended to gather information at the more useful planning scales of 1:50,000 or, exceptionally, 1:10,000.

The selected sites illustrated in Figure 7.2 cover both existing and planned infrastructural improvements, notably highways and hydroelectric schemes. Areas which lie physically remote from the main centres of communication have not been scheduled for development. This suggests that the government regards the ecological–economic zoning exercise as a tool for managing areas of existing development, rather than as a mechanism for extending development processes into new areas. While this is laudable, some concern remains that through its focus on the social and environmental effects of developments already under way, the zoning exercise will detract attention from the root causes. The development strategy currently pursued in Roraima requires reconsideration along fundamental lines, and it would be disappointing if the zoning exercise proved to be an excuse for failing to address two fundamental problems.

The first problem is that the development projects being implemented in Roraima are rarely tailored to the needs of the majority of the state's inhabitants. This is a fair criticism of an agricultural policy that expands colonisation schemes yet overlooks the requirements of the current smallholder community. Similarly, the government invests in large electrical generating plants for whose output there is little current nor foreseeable demand of any magnitude. Instead of improving the currently low standard of facilities for health and education, resources are applied to tax-free commerce that will mostly benefit external companies. All of these examples, as well as the rural to urban migration they stimulate, are indicative of a development philosophy that fails to identify and deal with basic human needs. The emphasis on large-scale infrastructural projects suggests that development continues to embody traditional and arguably outdated geopolitical strategies of frontier occupation.

The second problem concerns economic growth which is sustained on the

Figure 7.2 Priority areas for ecological-economic zoning of Roraima.

1 *Rio Cotingo Basin*: includes plans for an HEP installation; claims a high mineral, agro-pastoral and eco-tourist potential (especially in the Mount Roraima National Park); recognises that the area is one of social tension (ranching and agricultural expansion versus a strong Indian population).

2 *Caracaraí – Amazonas border (BR 174 corridor)*: includes plans for colonisation expansion and improvement of the road connection to Manaus; claims a timber, forest extraction (rubber, nuts) and eco-tourist (rain forest) potential.

3 *Confiança – Novo Paraíso road (RR 170)*: development of the road connection eliminating the ferry over the Rio Branco at Caracaraí; plans to expand agricultural colonies; claims timber and forest extraction potential; source of construction materials.

4 *Eastern limb of Perimetral Norte*: plans expansion of agricultural colonisation, timber and forest extraction; claims eco-tourist, high mineral and HEP (Jatapú and Anauá rivers) potential; recognises potential future link with Pará and Amapá.

5 *Middle and Lower Mucajaí Basin*: claims good prospects for agricultural colonisation (Apiaú and Caracaraí districts); hopes to develop high HEP potential (the unfinished Paredão site)

basis of rapid exploitation of natural resources. While mineral extraction necessarily depletes non-renewable deposits, current management practices in agriculture, ranching and logging are, in effect, constraining potential regeneration of renewable resources. The future consequences of resource depletion are given scant consideration by a development philosophy that seeks to maximise short-term profits. Furthermore, as it is the most marginalised rural communities that bear the brunt of environmental degradation, the subject receives little attention at state level where politics are orientated more towards the electorally significant urban population. The federal government organisation responsible for environmental affairs (IBAMA) appears to be fairly ineffective in Roraima, and so the law remains for the most part rhetorical. It is clear, however, that IBAMA is destined to have only a limited influence, as long as the powerful socio-economic and political forces that reward unsustainable exploitation of natural resources continue to dominate land development in the state.

The argument presented here is that the first stage in finding a solution for many of Roraima's environmental problems lies in the implementation of a more effective social policy. Securing land tenure for settlers and controlling urban migration within Roraima, as well as inward migration from other states, are prerequisites for improved management of land. Long-term management strategies are only likely to be successful with social stability. Migration from rural areas to the towns can be reduced by taking greater account of the views of rural people and incorporating them in the planning and management of development projects. This has been demonstrated by the work of the rubber tappers' council (Conselho Nacional de Seringueiros) in Acre. The involvement of local people is fundamental if development is to cater for their needs and if the processes that marginalise rural communities are to be reversed.

---

and captialise on proximity to mineral reserves (in the Yanomami Reserve – Serra de Surucucus).

6 *Boa Vista – Caracaraí corridor (BR 174)*: plans for agro-pastoral and agro-industrial development; partly dependent upon improved links to the south; looks to future link with Amazonas via São Gabriel de Cachoeira (the controversial western limb of the Perimetral Norte).

7 *Boa Vista – Venezuelan border (BR 174)*: plans to develop a free trade zone at Vila Pacaraima and looks to stronger links with Venezuela and the Caribbean; claims a good agro-pastoral and tourist potential.

8 *Boa Vista – Bonfim corridor (BR 401)*: plans to develop a free trade zone at Bonfim and to open up the Guyana links now that the road connection to Georgetown is complete; claims a good agro-pastoral potential (irrigated rice and livestock).

9 *Rio Amajari Basin*: plans to develop claimed high potential for livestock and arable farming on patches of Terra Roxa soils (Chapter 1); land available and not utilised by federal government agencies; claims high mineral (Tepequém and Serra de Pacaraima) and timber/forestry potential.

10 *Boa Vista urban area*: economic and political centre of state; plans to develop industrial and manufacturing sector; conscious of lack of mechanisms to develop transport, of environmental deterioration and poor use of urban land.

In the search for land uses that are adapted to Roraima's natural regions it would be helpful for planners to take account of the skilled management shown in indigenous strategies. The agricultural, hunting, fishing and extractive activities of Indian communities and *caboclos* (riverine dwellers), represent working models of how social and economic needs may be fulfilled without eroding the natural resource base. This is not to say that such strategies are a solution for all other rural communities, but that there are valuable precepts for utilising more of the natural resource without degrading biological diversity or disturbing environmental equilibrium.

LONGER-TERM DEVELOPMENT

This discussion forces us to question the long-term objectives of development in Roraima. Economic growth and natural resource exploitation are commonly defended as prerequisites for reducing poverty and increasing human welfare. However, there is little evidence that this is being achieved in Roraima, where the recent wave of spectacular mineral wealth was accompanied by increasing violence and disease. The developments in southern Pará have been similarly questioned by Schmink and Wood (1992), who point out that nutrition rates and health have declined in the area of their study, São Felix do Xingu, where a similar model of rural growth based on road construction has been postulated. Of particular note was a considerable increase in the child mortality rate, one of the fundamental indicators of social welfare, between 1978 and 1990.

Two characteristics of frontier development thwart the improvement of living standards in Roraima. Firstly, the capital generated from resource exploitation is very unevenly distributed and, secondly, a large proportion of this income is removed from the state and invested elsewhere. Both processes ensure that the majority of the migrant workforce receive only a fragment of the real income they generate. In the absence of capital investment, they survive only by expanding their agricultural and mining activities into new areas. On average, the Roraima workforce enjoys levels of income significantly higher than the north-east, from where most of the immigrants have moved. Because the migrant workforce derives virtually all of its capital from the exploitation of natural resources, it can only be viewed as a temporary phenomenon – not dissimilar to the flush of nutrients received by Amazonian forest soils following burning. As the environment is degraded over the long term, the people who depend upon it for their livelihood can only get poorer. Native societies are the first affected, but future impacts of the same processes will subsequently be borne by other social groups – particularly as the state's economy is so highly geared towards primary production.

Notwithstanding such observations, this book demonstrates that land development in Roraima is primarily motivated by a natural, if ill-considered desire to achieve maximimum economic growth. Public sector funding in the 1970s was substituted by increased flows of private capital throughout the 1980s to fuel an explosive phase of economic expansion. Even so, at the start of the 1990s, it is questionable whether consistent economic growth can be maintained. Attempts to expand the mineral sector directly conflict with indigenous land rights and require a level of support in Congress that is unlikely to be forthcoming. Increases in livestock and agricultural production demand either the application of expensive inputs or the expansion of land area, the repercussions of which are becoming increasingly sensitive. The perceived alternative is subsidised urban growth but, as Roraima lacks an industrial or manufacturing sector at present, this will probably only favour external companies who establish sub-divisions in the newly established Free Trade Zones. Whatever the obstacles that might prevent Roraima from becoming an economically viable state, none of these possibilities will necessarily improve the quality of life unless a greater effort is made by the state government to prevent capital flight and to redistribute the proceeds of development towards the sectors of society that need them most.

CONCLUSIONS: THE PAST, A POOR INDICATOR OF THE FUTURE?

Peter Furley

The case studies and their relationship to Amazonian development

Emphasis has been given to different aspects of the land development process, although it is evident that they are all closely related. The first two chapters set the work in context, tracing the natural resources and their relationship to economic and historical evolution. Chapter 1 revealed a predominantly forested region containing substantial and historically important tracts of savanna, both characterised by poor soils, generally good water supplies and rich but scattered mineral resources. The latter pose a dilemma, however, for they occur mostly on indigenous lands. The argument presented shows that Roraima has little political autonomy, yet, at the same time, its geographical location emphasises its isolation from the rest of Brazil. The state is characterised by rapid population growth, particularly concentrated in the capital, and a traditional settler reliance on ranching.

The complementary theme of Chapter 2 demonstrates the historical importance of the indigenous communities and illustrates the point that complex ecosystems have been used to support fairly large and complex societies without destroying the environment. Some, but certainly not all, Indian groups have survived contact and are increasing in numbers; others have retreated into a state of little contact, and have been seen to be very vulnerable – notably to the incursions of *garimpeiros*. Following through the history of settlement in Roraima underlines the dependence on communications with the rest of Brazil and the overwhelming role of ranching in shaping the character of the present state. The chapter recounts the history of social conflict between immigrant settlers and indigenous people, although the land degradation typical of areas south of the Amazon has not yet reached serious proportions in Roraima.

Subsequent chapters focus on the dynamics and strategies of rural development. Detailed evidence for the rates of deforestation and land-use change is illustrated in Chapter 3. Even in this remote part of the Amazon, between 2 to 4 per cent of the forest edge has been lost to clearance and a striking 25 per cent of the *várzea* disappeared between 1978 and 1985. The agricultural frontier is vividly observed gnawing away at the forest fringes; we can monitor this change but not, at present, control the causal processes. At a rancher level, Chapter 4 examines the pressures to clear forest and highlights the degradation that results. Continuous burning seems to result in a scrubby savanna, decreasing the buffering capacity of the vegetation against run-off, lowering rainfall absorption and increasing erosion and soil deterioration. At a smallholder level the same impelling processes are at work, and Chapter 5 shows how individual family strategies end up by clearing forest as the most favourable option. There are powerful environmental constraints at this level, and a highly seasonal agriculture in the northern parts of the state, leading to the adoption of multifaceted strategies for making a living. This accords with the pattern found in other parts of Amazonia and illustrates the difficulties facing policies which seek to intensify the use of abandoned land. The smallholders, who make up the bulk of the rural population, are shown to be a very mobile group. They are not easily managed, any more than ranchers or other land colonisers, in relation to existing environmental policies. In effect, a number of settlers form part of the informal economy and, to that extent, are out of direct governmental control.

The possibilities for arable and horticultural farming are explored in Chapter 6. There appears to be a more successful model for sustained farming here, based on the better soils and water supplies of the *várzea* lands and used at present for irrigated rice. Improving returns are evident, but there is an environmental cost in the loss of the protective riverine forest. These commercial operations favour large-scale, capital-intensive operations and clearly offer little prospect for smallholders. The provision of fruit and vegetables is examined along with the emergence of horticultural colonies specifically to cater for the only substantial urban market in Boa Vista. Clearly there is a growing demand directly related to the growth of population, but it is vulnerable, being dependent on so limited a market and on the problems of supply – coping with pests and diseases and the environmental constraints facing all forms of agricultural development in the region.

The concluding theme shows that the development model currently pursued in Roraima, if that is not altogether too grandiose an epithet for what has happened, is not solving any of the land development or social problems. It is a pattern of development that has been demonstrated elsewhere to be highly destructive to indigenous and some immigrant communities such as land squatters, whilst failing to provide an economic solution for sustaining the livelihood of the large majority of the inhabitants. Both the history of the gold-rush and the use of land by smallholders or ranchers are shown to have extensive social and environmental impacts. We therefore come to question whether it is appropriate to continue fostering development through land exploitation without addressing the external factors which most frequently determine the course of events.

Environment and development in a frontier state

Assuming that the government of Roraima is likely to pursue a model of rapid economic growth, then the present state of social inequality is likely to persist. Despite recent international attention on gold-mining and its impact on indigenous people, or criticisms of a national focus on the development of a tertiary sector through free trade zones, economic development in the near future is likely to depend upon some form of primary production. It is this purely economic solution to development which is questioned. On the one hand, gold, diamonds and cassiterite have made and may continue to make a major contribution to the economy of the state but frequently lie outside state control and, at the same time, there is little immediate prospect of a manufacturing or industrial growth. On the other hand, ranching, agriculture, agro-forestry and logging are proven if less spectacular providers of revenue. These could be expanded, but any disturbance to the natural vegetation has its environmental costs and it has proved exceptionally difficult to combine development with respect for the environment or the rights and wishes of the inhabitants.

There is much that could be done to improve agricultural productivity. A greater quantity of better quality and more diverse products could be generated from land already developed, but this cannot be achieved without greater investment. An expansion of the agricultural frontier might result from current research into agro-ecological zoning, but this will require a greater logic and justification than exists at present. The research and development objectives of EMBRAPA, currently useful but at a very small scale, and EMATER, the agricultural extension service, need greatly strengthening, as elsewhere in the Amazon. This is not only to help the commercial sector, such as irrigated rice, but should also improve the smallholder contribution. The aims of such strengthening would be firstly to raise production but, secondly, to avoid unnecessary soil, plant and water resource damage.

A better understanding of the natural resources is clearly needed, but is unlikely to be realised in the short term. The ecological and environmental zoning currently being undertaken in the Brazilian Amazon will add considerably to our knowledge of land potential. However, the sheer size of Roraima and the magnitude of the research necessary compared with the human resources available, means that the scale of the investigation is unlikely to be sufficiently detailed in the near future for satisfactory land-planning. Since Projeto Radambrasil published its reports some 20 years ago, little detailed research has been undertaken and a reconnaissance level knowledge of soils, vegetation and water resources and therefore land capability, is all that exists for a large part of the state. The current Rio Branco survey, part of the wider Ecological–Economic Zoning Plan, should provide more detailed data and a greater capability for analysing and interpreting resource information. In spite of international attention and the energetic activities of groups within Brazil such as the Church and various NGOs, the impacts of existing and potential developments on the environment, or for that matter on Roraima's inhabitants, have hardly been considered. What is on offer is a programme of more ranching, more land colonisation and some commercial agriculture and it is argued here that alternative strategies need to be developed. The current resource surveys should help to guide planning more effectively but, since mineral development faces severe constraints, the only innovative idea so far produced is that of free trade zones and it is arguable to what extent these will prove effective.

Political pressures and stresses are of particular importance in Roraima. The duality of federal and state control has been outlined, with notional ideas of more rational use of resources and conservation on the one hand and consequent difficulties for the state in developing a viable economy on the other. This question inevitably raises the crucial issue of self-sufficiency. The present assessment tends to be pessimistic: the state has control over only around a third of its territory; the soil resources are meagre and the limited area of soils which might sustain permanent agriculture requires constant inputs and enlightened, capital-intensive management; the water resources vary seasonally over a significant proportion of the area leading to water supply problems for irrigation and for the growing urban centres particularly around Boa Vista; rainfall in the savanna areas and forest–savanna margins is variable and unreliable and there are therefore, restricted possibilities for rain-fed agriculture in non-forested areas; mineral resources are unlikely to be used as the motor of economic development and, finally, the trappings of statehood demand a high level of raised revenue from a small population spread over a large area.

The state is unique in its geographical isolation from the centres of power in Brazil; in the individuality and strength of its indigenous peoples; in its high proportion of conserved and protected land; in its startling urban concentration in the capital, Boa Vista; and in its proximity to the rapidly developing southern regions of Venezuela. Nevertheless, numerous economic, political and social problems have still to be resolved. For the foreseeable future, exploitation of land resources is likely to remain the predominant economic force. After the surges in growth resulting from colonisation and then the gold-rush, the region waits on the prospect of a new stimulus.

NOTES

1 This controversial project which began in 1985, aims to strengthen external and internal security by delimiting a 'protection zone' from Amazonas to Amapá, covering an area which represents 14 per cent of Brazil. It involves increasing the military presence, clarifying frontiers, redefining the political handling of Indian people, road and hydroelectric construction, implantation of economic projects and colonisation schemes.

2 Letter from the Governor of Roraima to the President of Brazil, 17th October 1991, referring to Integração Fronteirica

3 SUDAM (1984: 37) notes that 'a indefinação vigente na demarcação das áreas de FUNAI tem descadeado graves focos de tensão social (em Roraima)'.

4 Speech given at the 'debate dos candidatos a governador/90' to the Sindicato dos Trabalhadores em Educação de Roraima (SINTER), 20 September 1990. Ottomar subsequently denied having made this remark when questioned at the SIMDAMAZONIA conference on Amazonian development held in Belém, February 1992.

5 Government estimate from Perfil Econômico do Estado de Roraima (1989); PRONASA, a group of extension workers who supervise livestock vaccinations in the state, estimate a 21 per cent decline in Roraima's beef herd from 354,000 in 1980 to 280,000 in 1989. Some experienced ranchers dispute this, arguing that the herd size in 1992 does not exceed 240,000 (pers. comm.).

6 Data from an unpublished report compiled by Catholic Church lawyers (January 1992), from a survey of ranches located within the proposed Makuxi reserve at Raposa/Serra do Sol.

7 Pers. comm. Dr Erasmus, specialist in the workings of the cassiterite market at DNPM, Manaus, November 1991.

8 'A brief rise in the price of tin in the first part of 1989 did not last because Brasil is able to place a further 10,000 metric tons or more originating from the *garimpo* of Surucucus and others in Roraima, on the market' (Brasil, DNPM 1990: 53).

9 The loan was made by the Midland Bank plc in 1984 and paid for the initial infrastructure including the access road and work on the hydroelectric plant, which has since been abandoned (Perfil Econômico do Estado de Roraima, 1989).

ACKNOWLEDGEMENTS

The contribution of the Royal Geographical Society through its Maracá Rain Forest Project is gratefully acknowledged for the earlier part of this work (PAF). Later research was funded through an ESRC studentship (GJM).

209

GLOSSARY

Agropecúaria	agro-pasture; arable plus livestock agriculture
Bandeirante	pioneer prospecting/exploring expeditions
Buritizal	line or clump of buriti palms (Mauritia flexuosa)
Caboclo	Amazonian backwoodsman; used to refer to racially mixed groups of immigrants with Indian blood and adopting Indian lifestyles
Campos	grasslands
Campos de murundus	field or tract of earthmounds
Campo sujo	grassland with scattered shrubs
Canteiro	small enclosure on raised bed
Capineira	pasture
Capoeira	fallow/cleared land
Capoeirão	long fallow/dense thicket
Cerrado	savanna (also sometimes used more technically to mean open arboreal/shrub savanna)
Colônia	colonial settlement; land colonisation scheme
Comarca	judiciary district of a state
Desenvolvimentista	someone advocating development/development philosophy
Farinha	flour or meal
Fazendeiro	owner of a fazenda – usually large-scale land holder
Filhos de criação	adopted 'godchild' of *fazendeiros*
Garimpagem	the search for mineral deposits; prospecting
Garimpo	mine; mining settlement
Igapó	inundated riverine woodland or swampland
Lavrado	ploughed land or land which has been worked
Maloca	Indian village
Mata ciliar	gallery forest
Mata de cipó	liane-dominated forest
Município	municipality; community with some self-government

210

Pantanal	swamp and wetland (after the area in southern Mato Grosso)
Perimetral Norte	Northern Perimeter highway
Picada	trail cut through forest
Posse	legal title to land; take-over of a stretch of land
Planalto	level upland; plateau
Proprietários dos regatões	river traders
Retiros	shelters on farmland
Sertanista	traveller to the outback or hinterland (the sertão)
Tepuis	isolated flat-topped mountains of the region, particularly in southern Venezuela
Terras devolutas	unoccupied lands
Titulo definitivo	clear title
Várzea	periodically or regularly flooded plains
Vazante	discharge or storm flow lines of a stream
Veredas	swampy depression/bottom land; often with a thin tree/shrub cover

BIBLIOGRAPHY

Abers, R. and Lourenço, A. (1992) 'Gold, geopolitics and hyperurbanisation in the Brazilian Amazon: the case of Boa Vista, Roraima', in G. Fadda (ed.), *La Urbe Latinoamericana ante el Nuevo Milenio*, Venezolana, Caracas: Fondo Editorial Acta Científica.

Adalbert, Dom O.S.B. (1913) 'Correspondance du Rio Branco', *Bulletin des Oeuvres et Missions Bénédictines au Brésil et au Congo*, Abbaye de Saint-André par Lophem, Belgium, 14 February.

Aguiar, Capt. Brás Diae de (1942) 'Trabalhos da Commisão Brasileira Demarcadora de Limites – Primeira Divisão – Nas Fronteiras da Venezuela e Guianas Britânica e Neerlandesa, de 1930 a 1940', *Anais IX Congresso Brasileiro de Geografia*, Rio de Janeiro.

Albert, B. (1992) 'Indian lands, environmental policy and military geopolitics in the development of the Brazilian Amazon: the case of the Yanomami', *Development and Change*, vol. 23, 35–70.

Allen, E. (1992) 'Calha Norte; military development in Brazilian Amazonia', *Development and Change*, 23, 71–99.

Almeida, A.L.S. de (1984) *Realizaçcões, Projeto Radambrasil*, Salvador, Bahia: Ministério das Minas e Energia.

Amodio, E. and Pira, V. (1986) 'Povos indigenas do Nordeste de Roraima', *Boletim*, vol. 11, February, Boa Vista.

Anderson, A.B. (1989) (ed.) *Alternatives to Deforestation*, New York: Columbia University Press.

——, and Anderson, S. (1983) *People and the Palm Forest*, Washington, DC: Man and the Biosphere Publication.

Anjos Filho, G.B. dos (1991) *Economia de Roraima – um breve perfil*, Boa Vista: SEBRAE-RR (Serviço de Apoio as Micro e Pequenas Empresas de Roraima).

Anon. (1984) 'Peru's Manu National Park in danger', *Wallaceana* 36 (11).

——, (1986) 'Sarney launches delayed program'. *Latin American Monitor*. (London) Mexico and Brazil, no. 3, 297–8.

Aubertin, C. (co-ord) (1988) *Fronteiras*, Brasília: Editora Universidade de Brasília.

Balik, M.J. (1982) 'Palmes neotropicales: nuevos fuentes de aceites comestibles', *Interciência* 7 (1), 25–9.

Barbira-Scazzochio, E. (ed.) (1980) *Land, People and Planning in Contemporary Amazônia*, Occasional Publications No. 3, Cambridge: Centre of Latin American Studies.

Barbosa, O. and Ramos, J.R.A. (1959) 'Território do rio Branco, aspectos principais da geomorfologia, da geologia e das possibilidades minerais de sua Zona Setentrional', *B. Div. Geol. Mineral*, no. 196, Rio de Janeiro.

Barrow, C.J. (1985) 'Development of the *várzeas* of Brazilian Amazonia, in J. Hemming (ed.), *Change in the Amazon Basin*, vol. 1: *Man's Impact on Forests and Rivers*, Manchester: University of Manchester Press.

——, (1990) 'Environmentally appropriate, sustainable small farm strategies for Amazonia', in D. Goodman and A. Hall (eds), *The future of Amazonia*, London: Macmillan.

Benchimol, S. (1988) *Amazônia Fiscal. Uma Analise da Arrecadação Tributaria e seus Efeitos sobre o Desenvolvimento Regional.* Manaus: Instituto Superior de Estudos da Amazônia.

Bigarella, J.J. and Ferreira, A.M.M. (1985) 'Amazonian geology and the Pleistocene and the Cenozoic environments and paleoclimates', in G.T. Prance and T.E. Lovejoy (eds), *Key Environments: Amazonia*, Oxford: Pergamon Press.

Binswanger, H.P. (1991) 'Brazilian policies that encourage deforestation in the Amazon', *World Development* 19(7), 821–9.

Black, C. A., Evans, D.D., White, J.L., Ensminger, L.E. and Clark, F.E. (1965) *Methods of Soil Analysis*, Madison, Wisconsin: American Society of Agronomy.

Bourne, R. (1978) *Assault on the Amazon*, London: Victor Gollancz.

Brand, J. (1988) 'The transformation of rainfall energy by a tropical rainforest canopy in relation to soil erosion' in P.A. Furley (ed.), *Biogeography and Development in the Humid Tropics*, Special Issue, *Journal of Biogeography*, vol. 15, 41–8.

Brasil (1975a) *Levantamento de Recursos Naturais*, Vol. 8, Folha NA. 20 Boa Vista e Parte das Folhas; NA.21 Tumucumaque; NB. 20 Roraima; NB.21 Departamento Nacional da Produção Mineral, Rio de Janeiro: Ministério das Minas e Energia.

——, (1975b) *Levantamento de Recursos Naturais*, Vol. 9, Folha NA.21 Tumucumaque e Parte da Folha NB.21; Departamento National da Produção Mineral, Rio de Janeiro: Ministério das Minas e Energia.

——, (1978) *Levantamento de Recursos Naturais*, Vol. 18, Folha SA.20 Manaus, Departamento Nacional da Produção Mineral, Rio de Janeiro: Ministério das Minas e Energia.

——, SUPLAN (1980) *Aptidão agrícola das terras de Roraima*, Secretaria Nacional de Planejamento Agrícola, Ministério of Agricultura, Brasília.

——, EMBRAPA (1981) *Cultivares de arroz de sequeiro para o Território Federal de Roraima*, Empresa Brasileiro Agropecuária, Circular Técnica no. 18 (April), Belém.

——, IBDF (1983) *Alteração da cobertura vegetal natural do Território de Roraima*, Brasília, Anexo Relatório Técnico, Instituto Brasileiro de Desenvolvimento Florestal, Ministério da Agricultura.

——, DNPM (1990) *Sumário Mineral 1987–1990*, Brasília. Departamento Nacional da Produção Mineral, Ministério das Minas e Energia.

——, SUDAM (1984) *O fenomeno migratório e a ação articulada dos programas de desenvolvimento de comunidade e migrações internas no território federal de Roraima*, Belém: Superintendência do Desenvolvimento da Amazônia.

——, SUDAM/OEA/PROVAM (1991) *Atualização dos estudos básicos do Vale do Rio Branco*, Belém: PR-SDR/SUDAM/SAP (June).

Buschbacher, R.J. (1986) 'Tropical deforestation and pasture development,' *BioScience*, vol. 36, 22–8.

——, Uhl, C. and Serrão, E.A. (1986) 'Pasture management and environmental effects near Paragominas, Pará', in C.F. Jordan (ed.), *Amazonian Rainforests, Ecosystem Disturbance and Recovery*, New York: Springer-Verlag.

Butt, Y. and Bogue, D. (1990) *International Amazonia: Its Human Side*, Chicago: Social Development Centre.

213

BIBLIOGRAPHY

Calzavara, B.B.G. (1972) *As possibilidades do açaizero no Estuario Amazônico*, FCAP Boletim No. 5, Belém: Faculdade de Ciências Agrárias do Estado do Pará.
CEDI (Centro Ecumênico de Documentação e Informação) (1985) *Aconteceu*, Annually for 1983 and subsequent years, São Paulo.
——, (1987) *Terras indigenas no Brasil*. São Paulo.
Cleary, D. (1987) *Anatomy of a Gold Rush*, Oxford: Oxford University Press.
——, (1991) *The Brazilian Rainforest; Politics, Finance, Mining and the Environment*, Special Report No. 2100, The Economist Intelligence Unit.
Cochrane, T.T. and Sanchez, P.A. (1982) 'Land resources, soils and their management in the Amazon', in *Amazonia, Agriculture and Land Use Research*, in S.B. Hecht (ed.), Cali, Columbia: Centro Internacional de Agricultura Tropical (CIAT).
——, ——, Azevedo, L.G. de, Porras, J.A. and Garver, C.L. (1985) *Land in Tropical America*. (3 vols), Cali, Colombia: Centro Internacional de Agricultura Tropical (CIAT).
Coelho, H.E. (1982) *Provárzeas: arroz irrigado*, Mimeo (29 April), Boa Vista.
Cordeiro, A.C.C. (1984) *Espaçamento e densidade de semeadura para o arroz em várzeas de Roraima*, Boa Vista: EMBRAPA/UEPAT, *Pesquisas em andamento*, (March).
——, and Mascarenhas, R.E.B. (1983) *Compartamento de cultivares de arroz em várzeas de Roraima*, Boa Vista: EMBRAPA/UEPAT, *Pesquisas em andamento*, no. 3 (February).
Costa, C. Nova da (1949) *O Vale do Rio Branco (suas realidades e perspectivas)*, Rio de Janeiro.
Coudreau, H.A. (1886) *Voyage au Rio Branco, aux Montagnes de la Lune, au Haut Trombettas*, Rouen, France.
Couto, W.S. and Alves, A.A.C. (1981) *Adubação mineral do arroz em solos de campo cerrado em Roraima*, Belém: EMBRAPA, *Pesquisas em andamento*, no. 47 (May).
Coy, M. (1987) 'Rondônia: frente pioneira e programa polonoroeste, O processo de diferenciação sócio-econômica na periferia e os limites do planejamento público', in G. Kohlep and A. Schrader (eds), *Homem e Natureza na Amazônia*, Tubingen: Universität Tubingen.
Cross, A. (1990) *Tropical Deforestation and Remote Sensing: The use of NOAA/ AVHRR Data over Amazonia*, UNEP/GRID, Geneva.
Cruz, S.A.S. (1980) *Garimpo do Tepequém: aspectos geológico e geoeconômico*, Boa Vista.
Dantas, M. and Rodrigues, I.A. (1980) 'Plantas invasoras de pastagens cultivadas na Amazônia', *Boletim de Pesquisa*, vol. 1, 1–23, EMBRAPA/CPATU, Belém.
Davis, S.H. (1977) *Victims of the Miracle, Development and the Indians of Brazil*, Cambridge: Cambridge University Press.
Detwiler, R.P. and Hall, C.A.S. (1988) 'Tropical forests and the global carbon cycle', *Science* 239, 42–7.
Dias, A.C.P. and Nortcliff, S. (1985) 'Effects of two land clearing methods on the physical properties of an Oxisol in the Brazilian Amazon', *Tropical Agriculture* 62, 207–12.
Dickinson, R.E. (1981) 'Effects of tropical deforestation on climate', *Studies in Third World Societies*, vol. 14, 411–41.
——, (1987) *Geophysiology of Amazonia: Vegetation and Climate Interactions*, New York: Wiley.
Diniz, E.S. (1972) *Os índios Macuxi de Roraima – sua instalação na sociedade nacional*, São Paulo.
Diniz, M. de Araujo Neto., Furley, P.A., Haridasan, M. and Johnson, C. (1986)

The murundus of the cerrado region of Central Brazil', *Journal of Tropical Ecology*, vol. 2, 17–35.

Dobson, A., Jolly, A. and Rubenstein, D. (1989) 'The greenhouse effect and biological diversity', *Trends in Ecology and Evolution*, vol. 4, 64–8.

Downs, G. (1991) 'Developing confidence levels for spatial operations in GIS: a case study,' Unpublished M.Sc. thesis, University of Edinburgh.

Eden, M.J. (1974) 'Palaeoclimatic influences and the development of savanna in southern Venezuela', *Journal of Biogeography*, vol. 1, 95–109.

——, (1978) 'Ecology, and land development: the case of Amazonian rainforest', *Transactions of the Institute of British Geographers*, New Series no. 3, 444–63.

——, (1982) 'Silvicultural and agroforestry developments in the Amazon basin of Brazil', *Commonwealth Forestry Review*, vol. 61, 195–202.

——, (1987) 'Traditional shifting cultivation and tropical forest system', *Trends in Ecology and Evolution*, vol. 2, 340–3.

——, (1990) *Ecology and Land Management in Amazonia*, London: Belhaven Press.

——, and Andrade, A. (1987) 'Ecological aspects of swidden cultivation among the Andoke and Witoto Indians of the Colombian Amazon', *Human Ecology*, vol. 15, 339–59.

——, and McGregor, D.F.M. (1989) 'Projeto Maracá 1987–88: Geography-Geografia' Unpublished manuscript and air photograph mosaic, Royal Botanic Garden, Edinburgh.

——, Furley, P.A., McGregor, D.F.M., Milliken, W. and Ratter, J.A. (1991) 'Effect of forest clearance and burning on soil properties in northern Roraima, Brazil', *Forest Ecology and Management*, vol. 38, 283–90.

Eiten, G. (1972) 'The cerrado vegetation of Brazil', *Botanical Review*, vol. 38, 201–341.

Environmental Defense Fund (EDF) (1991) *Brazilian Forest Policy in the Collor Government*, Washington, DC., June.

Falesi, I.C. (1976) 'Ecossistema de pastagem cultivada na Amazônia brasileira', EMBRAPA *Boletim Técnico*, Belém, vol. 1, 1–193.

Fearnside, P.M. (1979) 'Cattle yield prediction for the Transamazon highway of Brazil', *Interciência*, vol. 4, 220–5.

——, (1980) 'The effects of cattle pasture on soil fertility in the Brazilian Amazon: consequences for beef production sustainability', *Tropical Ecology*, vol. 21, 125–37.

——, (1985a) 'Environmental change and deforestation in the Brazilian Amazon, in J. Hemming (ed.), *Change in the Amazon Basin*, Vol. 1: *Man's Impact on Forests and Rivers*, Manchester: Manchester University Press.

——, (1985b) 'Brazil's Amazon forest and the global carbon problem,' *Interciência*, vol. 10, 179–96.

——, (1986a) *Human Carrying Capacity of the Brazilian Rainforest*, New York: Columbia University Press.

——, (1986b) 'Spatial concentration of deforestation in the Brazilian Amazon', *Ambio* 15(2), 74–81.

——, (1989) *A ocupação humana de Rondônia; impactos, limites e planejamento*, Programa Polonoroeste, Relatório de Pesquisa no. 5, SCT/PR CNPq. Assessoria Editorial e Divulgação Científica, Brasília.

——, and Ferreira, G. de. L. (1984) 'Roads in Rondônia: highway construction and the farce of unprotected forest reserves in Brazil's Amazonian forest', *Environmental Conservation*, vol. 11, 358–60.

——, (1990) 'Environmental destruction in the Brazilian Amazon,' in D. Goodman and A. Hall (eds), *The Future of Amazonia*, London: Macmillan.

Fearnside, P.M. Tardin, A.T. and Meira Filho, L.G. (1990) *Deforestation Rate in Brazilian Amazonia*, Brasília: National Secretariat of Science and Technology.

Feijão, A.J. and Pinto, J.A. (1990) *Garimpeiros Activities in South America: the Amazon Gold Rush*, USAGAL/BMF (União dos Sindicatos Associações de Garimpeiros da Amazônia Legal/Bolsa Mercantil de Futuros, Pará.

Fernandes, F.R.C. and I.C. de M. Portela (1991) *Recursos minerais da Amazônia – algunos dados sobre situação e perspectivas*, CETEN – CNPq (Conselho Nacional de Pesquisa).

Ferreira, J.P. (1957) *Encyclopédia dos Municípios Brasileiros* (14 vols), Rio de Janeiro.

Ferri, P. (1990) *Achados ou perdidos? A imigração indigena em Boa Vista*, MLAL, Goiânia, Goias.

Figueiredo, L. (1944) 'Fronteiras amazônicas', in IBGE, *Amazônia Brasileira*, Rio de Janeiro: IBGE.

Forsberg, B. Godoy, J.M., Victoria, R. and Martinelli, L.A. (1989) 'Development and erosion in the Brazilian Amazon: a geochronological study', *GeoJournal*, vol. 19, 402–5.

Fortunato, J.M.P. and Pinto F.D. de J.M. (1991) *Atualização dos estados básicos do Vale do Rio Branco*, Organisação do Espaço 2° versão, OEA/PROVAM, PRSDR/SUDAM/SAP, Belém.

Fränzle, O. (1979) 'The water balance of the tropical rain forest of Amazonia and the effects of human impact', *Applied Sciences and Development*, vol. 13, 88–107.

Freitas, L.A. Soares de (1990) *A historia política e administrativa de Roraima 1945–85* (ed. Palmares), Porto Velho, Rondônia.

—— , (1991) *Politicas públicas e administrativas de Territórios Federais do Brasil*, Boa Vista: Editôra Boa Vista.

Furley, P.A. (1986) 'Radar surveys for resource evaluation in Brazil: an illustration from Rondônia', in M.J. Eden and J.T. Parry (eds), *Remote Sensing and Tropical Land Management*, Chichester: Wiley.

—— (1990) 'The nature and sustainabilty of Brazilian Amazon Soils', in D. Goodman and A. Hall (eds), *The Future of Amazonia*, London: Macmillan.

—— (in 1993) 'Tropical moist forests: transformation or conservation?', in N. Roberts (ed.), *The Changing Global Environment*, Oxford: Basil Blackwell.

—— and Ratter, J.A. (1988) 'Soil resources and plant communities of the central Brazilian cerrado and their development', in P.A. Furley, *Biogeography and Development in the Humid Tropics*, Special Issue, *Journal of Biogeography* 15(1), 97–108.

—— , —— (1990) 'Pedological and botanical variations across the forest–savanna transition on Maracá Island', *Geographical Journal* 156(3), 251–66.

—— , Proctor, J. and Ratter, J.A. (eds) (1992) *The nature and Dynamics of Forest–Savanna Boundaries*, London: Chapman & Hall.

—— , Dargie, T.C.D. and Place, C. (1993) 'Remote sensing and the establishment of a GIS for resource management on and around Maracá Island,' in J. Hemming (ed.), *The Rain Forest Edge*, Manchester: Manchester University Press.

Gentry, A.H. and López-Parodi, J. (1980) 'Deforestation and increased flooding of the upper Amazon', *Science*, no. 210, 1354–6.

Gianluppi, V., Moraes, E. de and Camargo, A.H.A. (1983) *Sistema de produção sequencial de arroz com ferrageiras, em solos de cerrado de Roraima; 1. Andropogon e Guandu*, EMBRAPA/UEPAT *Pesquisas em andamento*, no. 4 (February), Boa Vista.

Goedert, W. (1983) 'Management of cerrado soils of Brazil: a review,' *Journal of Soil Science* 34(3), 405–28.

Goldstein, I.S. (1979) 'Chemicals from wood', *Unasylva*, vol. 31, 1252–9.
Gómez-Pompa, A., Vásquez-Yanes, C. and Guevara, S. (1972) 'The tropical rain forest: a non-renewable resource,' *Science*, no. 177, 762–5.
——, Whitmore, T.C. and Hadley, M. (eds) (1991) *Rain Forest Regeneration and Management*, Carnforth, Lancashire: UNESCO-Parthenon Publishing.
Gondim, J. (1922) *Através do Amazonas*, Manaus.
Goodland, R. J. A. (1980) 'Environmental ranking of Amazonian development projects in Brazil', *Environmental Conservation* 7(1), 9–26.
—— and Irwin, M.S. (1975) *Amazon Jungle, Green Hell to Red Desert?*, Amsterdam: Elsevier.
——, —— and Tillman, G. (1978) 'Ecological development for Amazonia', *Ciência e Cultura* 30(3) 275–89.
—— and Ferri, M. (1979) *Ecologia do cerrado*, São Paulo: Editôra Universidade.
Goodman, D. and Hall, A. (eds) (1990) *The Future of Amazonia: Destruction or Sustainable Development?*, London: Macmillan.
Gradwohl, J. and Greenberg, R. (1988) *Saving the Tropical Forests*, London: Earthscan Publications.
Guerra, A.T. (1957) *Estudo geográfico do Território Federal do Rio Branco*, Rio de Janeiro.
Gusmão, S. de (n.d.) *Riquezas e segredos da Amazônia*, Manaus.
Haffer, J. (1969) 'Speciation in Amazonian forest birds', *Science*, no. 165, 131–7.
Hall, A. (1989) *Developing Amazonia: Deforestation and Social Conflict in Brazil's Carajás Programme*, Manchester: Manchester University Press.
Hamilton Rice, A. (1928) 'The Rio Branco, Uraricuera, and Parima', *The Geographical Journal* 71(2) 113–43 and (3), 209–29, 343–57.
Hare, F.K. (1980) 'The planetary environment: fragile or sturdy?', *Geographical Journal*, no. 146, 379–95.
Harris, D.R. (1971) 'The ecology of swidden cultivation in the upper Orinoco rain forest, Venezuela,' *Geographical Journal*, no. 61, 475–95.
Hecht, S.B. (1981) 'Deforestation in the Amazon basin: magnitude, dynamics and soil resources effects', *Studies in Third World Societies*, vol. 13, 61–108.
—— (1985) 'Environment, development and politics: capital accumulation and the livestock sector in eastern Amazonia', *World Development*, vol. 13, 663–84.
——, and Cockburn, A. (1989) *The Fate of the Forest*, London: Verso.
——, Norgaard, R. and Possio, G. (1988) 'The economics of cattle ranching in eastern Amazonia', *Interciência* 13(5), 233–40.
Hemming, J. (1978) *The Search for El Dorado*, London: Michael Joseph.
—— (ed.) (1985) *Change in the Amazon Basin*, (2 vols), Manchester: Manchester University Press.
—— (1987) *Amazon Frontier. The defeat of the Brazilian Indians*, Cambridge, Mass.: Harvard University Press.
—— (1988) 'Denizens of the rainforest', *Geographical Magazine* LX(8), 2–10.
—— (1989) 'A expedição ecológica de Maracá,' *Horizonte Geográfico* 2(6), 16–32.
—— (1990a) *Roraima: Brazil's Northernmost Frontier*, Research Paper 20, London: Institute of Latin American Studies.
—— (1990b) 'How Brazil acquired Roraima', *The Hispanic American Historical Review* 70(2), 295–325.
—— (ed.) (1993) *The Rain Forest Edge*, Manchester: Manchester University Press.
—— and Ratter, J.A. (1993) *Maracá – Rainforest Island*, London: Macmillan.
HMSO (1991) *Climatological and Environmental Effects of Rainforest Destruction*, Environmental Committee 3rd Report, London.

Holden, C. (1979) 'Park is sought to save Indian tribe in Brazil', *Science*, no. 206, 1160–2.

Hopper, J. H. (1966) (ed.) *Indians of Brazil in the Twentieth Century*, Washington, DC.

IBGE (1944) *Amazônia Brasileira*, Rio de Janeiro: Fundação Instituto Brasileiro de Geografia e Estatística.

——, (1976) *Sinopse estatística Roraima 1975*, Rio de Janeiro.

——, (1981) *Atlas de Roraima*, Rio de Janeiro: Fundação Instituto Brasileiro de Geografia e Estatística.

——, (1983) 'Censo Agropecuário: Roraima-Amapá', *IX Recenseamento Geral do Brasil 1980* 2–3 (5), Rio de Janeiro.

—— (1986) *Anuario Estatístico do Brasil*, Rio de Janeiro.

—— (1987) 'Sinopse preliminar do censo agropecuário', *Região Norte, Censos econômicos – 1985* 4(1), Rio de Janeiro: Região Norte.

—— (1988) *Anuario Estatístico do Brasil 1987*, Rio de Janeiro.

—— (1992) *Roraima: Censo Demográfico 1991, Resultados Preliminares, Janeiro 1992*, Boa Vista.

IDB/UNDP/TCA (1992) *Amazonia Without Myths*, Washington, DC: Commission on Development and Environment for Amazonia, Interamerican Development Bank/United Nations Development Programme/Amazon Co-operation Treaty.

Jatene, H. da Silva (co-ord) (1984) *O fenomeno migratório e ação articulada dos programas de desenvolvimento de comunidade e migrações internas no Território Federal de Roraima*, Belém: Departamento de Recursos Humanos, SUDAM (Superintendência de Desenvolvimento da Amazônia).

Jones, J.R. (1990) *Colonisation and Environment: Land Settlement Projects in Central America*, Tokyo: United Nations University.

Jordan, C.F. (1985) *Nutrient Cycling in Tropical Forest Ecosystems*, Chichester: Wiley.

—— (ed.) (1987) *Amazonian Rain Forests*, Ecological Studies 60, New York: Springer-Verlag.

Joyce, C. (1985) 'Tapping the lungs of the earth', *New Scientist*, no. 1466, 48–50.

Kelsey, T.F. (1972) 'The beef cattle industry in the Roraima savannas: a potential supply for Brazil's north', Doctoral dissertation, University of Florida.

Kietman, D. (1966) 'Indians and culture areas of twentieth century Brazil', in J.H. Hopper (ed.), *Indians of Brazil in the Twentieth Century*, Washington DC.

Koch-Grünberg, T. (1913) 'Meine Reise durch Nord-Brasilien zum Orinoco, 1911–1913', *Zeitschrift der Gesellschaft für Erdkunde zu Berlin*, pp. 665–332.

Kubitzki, K. (1985) 'The dispersal of forest plants', in G.T. Prance and T.E. Lovejoy (eds), in *Amazonia: Key Environments*, Oxford: Pergamon Press.

Landon, J.R. (1984) *Booker Tropical Soil Manual*, Harlow: Longman.

Leite, L.L. and Furley, P.A. (1982) 'Land development in the Brazilian Amazon with particular reference to Rondônia and the Ouro Preto Colonisation Project', in J. Hemming (ed.), *Change in the Amazon Basin*, vol. 2, Manchester: Manchester University Press.

Léna, P. (1988) 'Diversidade da fronteira agrícola na Amazônia', in C. Aubertin (co-ord), *Fronteiras*, Brasília: Editôra Universidade de Brasília.

—— (forthcoming) 'Estrategia camponesa de capitalização no PIC Ouro Preto, Rondônia', in J. Hebette (ed.), *O cerco está se fechando?*, São Paulo: HUCITEC.

Lettau, H., Lettau, K. and Molion, L.C.B. (1979) 'Amazonia's hydrologic cycle and the role of atmospheric recycling in assessing deforestation effects,' *Monthly Weather Review*, no. 107, 227–8.

Lewis, G.P. and Owen, P.E. (1989) *Legumes of the Ilha de Maracá*, Royal Botanic Gardens, Kew.

Lima, A. (1944) 'A explotação amazônica', in IBGE, Amazônia brasileira, Rio de Janeiro: IBGE.

Lima, S.J. de (co-ord) (1984) Roraima: agricultura e sociedade, Diagnostico Geral do Setor Agrícola do Território Federal de Roraima, CEPA (Comissão de Planejamento Agricola, Boa Vista.

Lovejoy, T.E. and Padua, M.T. de J. (1980) Can Science Save Amazonia? London: Earthscan.

Maciel da Silveira, I. and Gatti, M. (1988) 'Notas sobre a ocupação de Roraima, migração e colonização,' Boletim do Museu Paraense Emílio Goeldi Antropologia 4(1), 43–64.

McConnell, R.B. (1968) 'Planation surfaces in Guyana', Geographical Journal, vol. 134, 506–520

McGregor, D.F.M. (1980) 'An investigation of soil erosion in the Colombian rainforest zone', Catena, vol. 7, 265–73.

—— and Eden, M.J. (forthcoming) 'Geomorphology of the Ilha de Maracá', in W. Milliken and J.A. Ratter, (eds), The Ecology of an Amazonian Rain Forest, Manchester: Manchester University Press.

MacMillan, G. (1993a) 'Ouro e Agricultura na Amazônia', Ciência Hoje, no. 87.

—— (1993b) Gold mining and land use in the Brazilian Amazon, unpublished Ph.D. thesis, University of Edinburgh.

McNeil, M. (1964) 'Laterite soils', Scientific American 211(5), 96–102.

Margolis, M. (1973) The Moving Frontier, Gainesville: University of Florida Press.

—— (1988) 'Battle brews over rush to Brazilian El Dorado', The Times, London, 5 March.

Marlier, G. (1973) 'Limnology of the Congo and Amazon Rivers', in Tropical Forest Ecosystems in Africa and South America: A Comparative Review, B.J. Meggers, E.S. Ayensu and W.D. Duckworth (eds), Washington DC: Smithsonian Institution.

Marques, J., Santos, J.M., dos, Villa Nova, N.A. and Salati, E. (1977) 'Precipitable water and water vapour flux between Belém and Manaus, Acta Amazônica, vol. 7, 355–62.

Martins, P.F. da S., Cerri, C.C., Volkoff, B., Andreux, F. and Chauvel, A. (1991) 'Consequences of clearing and tillage on the soil of a natural Amazonian ecosystem,' Forest Ecology and Management, vol. 38, 272–82.

Maslen, J. (1992) 'The design of a GIS for the management of ecological data on Tropical Forest Reserves', Unpublished M.Sc. University of Edinburgh.

May, R.M. (1975) 'The tropical rainforest', Nature, no. 257, 737–8.

Meggers, B.J. (1971) Amazonia: Man and Culture in a Counterfeit Paradise, Chicago: Aldine Atherton.

Migliazza, E. (1978) The Integration of the Indigenous Peoples of the Territory of Roraima, Brazil. IWGIA Doct. 32. Copenhagen.

Miller, D. (1985) 'Replacement of traditional elites: an Amazon case study,' in J. Hemming (ed.), Change in the Amazon Basin, Vol. 2: The Frontier after a Decade of Colonisation, Manchester: Manchester University Press.

Milliken, W. and Ratter, J.A. (1989) The Vegetation of the Ilha de Maracá, 1st Report of the Vegetation Survey of the Maracá Rain Forest Project. RGS/INPA/SEMA, Royal Botanic Garden, Edinburgh.

——, —— (eds) (forthcoming) The Ecology of an Amazonian Rain Forest, Manchester: Manchester University Press.

Molion, L.C.B. (1976) A Climatonomic Study of the Energy and Moisture Fluxes of the Amazonas Basin with Considerations of Deforestation Effects, INPE (Instituto de Pesquisas Espaciais), São José dos Campos, SP, Brazil.

Mollard, M. (1913) *Le Haut Rio Branco, Région d'alimentation de tout le basin de l'Amazone*, Rio de Janeiro.

Moran, E.F. (1981) *Developing the Amazon*, Bloomington: Indiana University Press.

—— (1983) *The Dilemma of Amazonian Development*, Boulder, Colo.: Westview Press.

——, (1990) 'Private and public colonisation schemes in Amazonia', in D. Goodman and A. Hall (eds), *The Future of Amazonia*, London: Macmillan.

Morgan, R.P.C. (1986) *Soil Erosion and Conservation*, Harlow: Longman.

Mors, W.B. and Rizzini, C.T. (1966) *Useful Plants of Brazil*, San Francisco: Holden-Day.

Mougeot, L.J.A. (1983) 'A retenção migratória das cidades pequenas nas frentes, amazônicas de expansão: um modêlo interpretativo', in L.J.A. Mougeot and L.E. Aragon (eds), *O despovoamento do território amazônico; contribuições para sua interpretação*, Belém: Gráfica Falangola Editôra Ltda.

Myers, I. (1945–46) 'The Makushi of British Guiana – a study in culture and contact', *Timehri: Journal of the Royal Agricultural and Commercial Society of British Guiana*, vol. 26, 66–77; vol. 27, 16–38.

——, (1988) 'The Makushi of the Guiana–Brazilian frontier in 1944: a study of cultural contact', manuscript.

Myers, N. (1977) 'The nature of the deforestation problem – trends and policy implications,' in D.R. Shane (ed.), *Proceedings of the U.S. Strategy Conference on Tropical Deforestation*, US Department of State/US Agency for International Development, Washington, DC.

—— (1979) *The Sinking Ark. A New Look at the Problem of Disappearing Species*, Oxford: Pergamon Press.

—— (1980) *Conversion of Tropical Moist Forests*, Washington, DC: National Academy of Sciences.

—— (1984) *The Primary Source: Tropical Forests and Our Future*, New York: W.W. Norton.

—— (1986) 'Tropical deforestation and a mega-extinction spasm,' in Soulé (ed.), *Conservation Biology: the Science of Scarcity and Diversity*, Sunderland: Sinauer Associates.

—— (1988) 'Tropical forests: much more than stocks of wood', *Journal of Tropical Ecology*, 4, 209–221.

Nabuco, J. (1903) *Limites entre le Brésil et la Guyane Anglaise, Question soumise à l'arbitrage de S.M. Le Roi d'Italie* (8 vols), Rio de Janeiro.

National Academy of Sciences (1975) *Underexploited Tropical Plants with Promising Economic Value*, Washington, DC.

Nelson, R., Horning, N. and Stone, T.A. (1987) 'Determining the rate of forest conversion in Mato Grosso, Brazil, using Landsat MSS and AVHRR data', *International Journal of Remote Sensing*, vol. 8, 1767–84.

Neto, R.B. (1989) 'Disputes about destruction', *Nature*, no. 338, p. 531.

Nicholaides, J.J. III, Bandy, D.E., Sanchez, P.A., Villachia, J.H., Couto, A.J. and Valverde, C.S. (1984) 'Continuous cropping potential in the upper Amazon basin,' in M. Schmink and C.H. Wood (eds), *Frontier Expansion in Amazonia*, Gainesville: University of Florida Press.

Nisbet, E.G. (1988) 'The business of plant management', *Nature*, no. 333, p. 617.

Norden, S.F. and Meade, R.H. (1982) 'Deforestation and increased flooding of the upper Amazon', *Science*, no. 215, 426–7.

Nortcliff, S. and Robison, D. (1988) 'The soils and geomorphology of the Ilha de

Maracá, Roraima, second approximation', Department of Soil Science, Reading (manuscript).
——, Ross, S.M. and Thornes, J.B. (1989) 'Soil moisture runoff and sediment yield from differentially cleared tropical rainforest plots', in J.B. Thornes (ed.), *Vegetation and Erosion*, Chichester: Wiley.
Odum, E.P. and Odum, H.T. (1972) 'Natural areas as necessary complements of man's total environment', *Transactions of the 37th North American Wildlife and Natural Resources Conference*, Mexico City, March, pp. 178–89.
Oliveira, A. and Furley, P.A. (1990) 'Monchão, Cocurutu, Murundu', *Ciência Hoje* 11(61), 30–7.
Oliveira, A. I. de (1929) *Bacia do Rio Branco (Estado de Amazonas)*, Serviço Geológico e Mineralógico do Brasil, no. 37, Rio de Janeiro.
—— (1937) 'Recursos minerais da bacia do Rio Branco', *Miner Metal* (Rio de Janeiro) 1(6), 243–350.
Oliveira, I.W.B. *et al.* (1969) *Nota sobre a geologia e recursos minerais da área do Projeto Roraima*, Belém, D.N.P.M.
Ourique, A.E.J. (1906) *O Valle do Rio Branco*, Manaus.
Pacheco, J. de Oliveira Filho (1990) 'Frontier security and the new indigenism: nature and origins of the Calha Norte Project', in D. Goodman and A. Hall (eds), *The Future of Amazonia*, London: Macmillan.
Paixão, E. Silva, M. (1943) *Sobre uma geografia social da Amazônia*, Manaus.
Pandolfo, C. (1974) *Estudos básicos para o Estabelecimento de uma Politica de Desenvolvimento dos Recursos Florestais e de Uso Racional das Terras da Amazônia*, Belém: SUDAM (Superintendência do Desenvolvimento da Amazônia).
Pearman, G.I. and Fraser, P.J. (1988) 'Sources of increased methane,' *Nature*, no. 332, 489–90.
Pereira, E.S., Schwade, E. *et al.* (1983) *Resistência Waimiri/Atroari*, Marewa, Itacoatiara.
Pereira, L. (1917) *O Rio Branco – Observações de viagem*, Manaus.
Phillips, S. (1990). 'The effects of forest clearance on the properties of soils from selected sites in Roraima, Brazil,' Unpublished ms, University of Edinburgh.
Polunin, N. (ed.) (1980) *Growth with Ecodisasters?*, London: Macmillan.
Ponce, V.M. and Cunha, C.N. da (forthcoming), 'Vegetated earthmounds in tropical savannas of central Brazil: a synthesis', Submitted to *Journal of Biogeography*.
Potter, G.L., Ellsaesser, H.W., MacCracken, M.C. and Luther, F.M. (1975) 'Possible climatic impact of tropical deforestation', *Nature*, no. 258, 697–8.
Prance, G.T. (1977) 'The phytogeographic subdivisions of Amazonia and their influence on the selection of biological reserves', in G.T. Prance and T.S. Elias (eds), *Extinction is Forever*, New York Botanical Garden, Bronx.
——, (ed.) (1982) *Biological diversification in the Tropics*, New York: Columbia University Press.
——, (1985) 'The changing forests', in G.T. Prance and T.E. Lovejoy (eds), *Key environments: Amazonia*, Oxford: Pergamon Press.
——, (1986) *Tropical Rain Forests and the World Atmosphere*, AAAS Selected Symposia No. 101, Boulder, Colorado: Westview Press.
——, (1990) 'Concensus for conservation', *Nature*, no. 345, p. 384.
——, and Lovejoy, T.E. (eds) (1985) *Key Environments: Amazonia*, Oxford: Pergamon Press.
Ramgrab, G.E., Bonfim, L.F.C. and Mandetta, P. (1972) *Projeto Roraima, 2ª fase*, Manus: DNPM (Departamento Nacional de Produção Mineral).
Ramos, A.R. and Taylor, K.I. (1979) *The Yanomama in Brazil*, IWGIA Document 37, Copenhagen.

Rangel, P.H.N., Galvão, E.U.P., Nogueira, O.L.N. and Behnck, A. (1978) *Avaliação de cultivares de arroz no Território Federal de Roraima*, EMBRAPA Communicado Técnico no. 4 (March), Boa Vista.

Ratter, J.A., Askew, G.P., Montgomery, R.F. and Gifford, D.R. (1978) 'Observations on forests of some mesotrophic soils in central Brazil', *Revista Brasileira Botânica*, vol. 1, 47–58.

Reis Altschul, S. von. (1977) 'Exploring the herbarium', *Scientific American* 236(5), 96–104.

Ribeiro, D. (1957) 'Culturas e linguas indigenas do Brasil', *Educação e Ciências Sociais*, vol. 2, Rio de Janeiro.

Ribeiro, N. de O. (ed.) (1969) *Território Federal de Roraima* (2 vols), Fundação Delmiro Gouveia, ms Rio de Janeiro (n.d. but around 1969).

Ribeiro, R.H.E., Cordeiro, A.C.C. and Rangel, P.H.N. (1985) *EMBRAPA/UEPAT, Pesquisas em andamento*, no. 1 (February), Boa Vista.

Richards, P.W. (1977) 'Tropical forests and woodlands: an overview', *Agro-Ecosystems*, vol. 3, 225–38.

Richey, J.E., Brock, J.T., Maiman, R.J., Wissmar, R.C. and Stallard, R.F. (1980) 'Organic carbon: oxidation and transport in the Amazon river', *Science*, no. 207, 1348–51.

Rivière, P. (1972) *The Forgotten Frontier. Ranchers of Northern Brazil*, New York: Holt, Rinehart & Winston.

Rodrigues, J.B. (1882) *Rio Jauapery: Pacificação dos Crichanas*, Rio de Janeiro.

Rondon, C.M. da Silva (1955) *Indios do Brasil*, SPI Publication 89, No. 3, Rio de Janeiro.

Roraima, ASTER-RR (1982) *Cultura do arroz em Roraima*, ASTER-RR/CEPA-RR, Boa Vista.

——, CEPA (1983) *Diagnóstico da situação atual da horticultura em Roraima*, Boa Vista: Comissão de Planejamento Agricola.

——, —— (1984) *Roraima: agricultura e sociedade, Diagnóstico Geral do Setor Agrícola do Território Federal de Roraima*, Secretaria de Agricultura, Comissão de Planejamento Agrícola, Convênio 1984, Boa Vista.

——, EMATER/EMBRAPA (1981) *Sistemas de produção para arroz, milho, mandioca, caupi e banana*, Sistemas de Produção, Boletim no. 375, Boa Vista.

——, EMBRAPA (1981) *Cultivares de arroz de sequeiro para o Território Federal de Roraima*, EMBRAPA Circular Técnica 18 (April), Belém.

——, EMBRAPA/EMATER (1977) *Sistemas de produção para arroz, mandioca e milho*, ASTER Sistemas de Produção, Boletim no. 71, Boa Vista.

——, —— (1979) *Sistemas de produção para o arroz, Território Federal de Roraima*, Sistemas de Produção, Boletim no. 165, Boa Vista.

——, SEAGRI-RR (1979) *Projeto Milho*, Secretaria de Agricultura, Boa Vista.

——, SEPLAC (1980) *Informações Estatísticas*, vol. 1, nos 1–2, Secretaria de Planejamento e Coordenação, Boa Vista.

——, SEPLAN-RR (n.d.) *Diagnóstico Fundiario – Roraima, Subsídios ao sistema Fundiario Nacional*, Secretaria de Planejamento, Boa Vista.

——, —— (1989) *Perfil econômico do Estado de Roraima*, Boa Vista.

——, SEPLAN (1991) *Programa especial de investimentos 1991–95*, Secretaria de Planejamento, Industria e Comercio, Boa Vista.

Ross, S. (1992) 'Soil and litter nutrient losses in forest clearings close to a forest–savanna boundary on Maracá Island, Roraima, Brazil', in P.A. Furley, J. Proctor and J.A. Ratter (eds), *Nature and Dynamics of Forest–Savanna Boundaries*, London: Chapman and Hall.

——, Thornes, J.B. and Nortcliff, S. (1990) 'Soil hydrology, nutrient and erosional response to the clearance of terra firme forest, Maracá Island, Roraima, Northern Brazil', *Geographical Journal* 156, 267–82.

——, Luizão, F.J. and Luizão, R.C.C. (1992) 'Soil conditions and soil biology in different habitats across a forest–savanna boundary on Maracá Island, Brazil', in P.A. Furley, J. Proctor and J.A. Ratter (eds), *Nature and Dynamics of Forest–Savanna Boundaries*, London: Chapman and Hall.

Sagan, C., Toon, O.B. and Pollack, J.B. (1979) 'Anthropogenic albedo changes and the earth's climate', *Science*, no. 206, 1363–8.

Salati, E., Olio, A.D., Matsui, E. and Gat, J.R. (1979) 'Recycling of water in the Amazon basin: an isotopic study', *Water Resources Research*, vol. 15, 1250–8.

——, Vose, P.B. and Lovejoy, T.E. (1986) 'Amazon rainfall, potential effects of deforestation, and plans for future research', in G.T. Prance (ed.) *Tropical Rain Forests and the World Atmosphere*, Boulder, Colo.: Westview Press.

Sanchez, P.A. (1976) *Properties and Management of Soils in the tropics*, New York: Wiley.

Santana, S.C. de (1986) *Arroz irrigado: Roraima já mostra sua força agrícola*, Extensão Rural (January–February), EMBRATER, Brasília, pp. 16–17.

Santilli, M. (1990) 'Projet Calha Norte: politique indiginiste et frontières nord-amazoniennes', in B. Albert (ed.), *Brésil: indiens et développement en Amazonie*, Revue de Survival International (France), Paris: Survival International.

Sarmiento, G. (1984) *The Ecology of Neotropical Savannas* (trans. O. Solbrig), Cambridge, Mass: Harvard University Press.

——, (1992) 'A conceptual model relating environmental factors and vegetation formations in the lowlands of tropical South America', in P.A. Furley, J. Proctor and J.A. Ratter (eds), *Nature and Dynamics of Forest–Savanna Boundaries*, London: Chapman and Hall.

Schmink, M. and Wood, C.H. (1985) *Frontier Expansion in Amazonia*, Gainesville, Fla.: University of Florida Press.

——, —— (1992) *Contested Frontiers in Amazonia*, New York: Columbia University Press.

Schomburgk, R.H. (1840a) 'Report of the Third Expedition into the interior of Guyana, comprising the journey to the sources of the Essequibo, to the Caruma Mountains, and to Fort San Joaquim on the Rio Branco, in 1837–38', *Journal of the Royal Geographical Society*, vol. 10, 159–90.

——, (1840b) 'Journey from Fort St. Joaquim on the Rio Branco to Roraima, and thence by the rivers Parima and Merewari to Esmeralda on the Orinoco in 1838–9', *Journal of the Royal Geographical Society*, vol. 10, 191–247.

Schwarz, W. and Rocha, J. (1989) 'Brazilian police and troops to protect Amazon forest', *The Guardian*, London, 15 February.

Seiler, W. and Conrad, R. (1987) 'Contribution of tropical ecosystems to the global budgets of trace gases, especially CH_4, H_2, CO and N_2O, in R.E. Dickinson (ed.), *The Geophysiology of Amazonia: Vegetation and Climate Interactions*, New York: Wiley & Sons.

SEMA (1977) *Program of Ecological Stations*, Brasília: Secretaria Especial do Meio-Ambiente.

SEMTUR (1986) *Município de Boa Vista*, Secretaria de Planejamento e Coordenação, Governo de Roraima, Boa Vista (Mimeo).

Serrão, E.A. and Toledo, J.M. (1988) 'Sustaining pasture-based production systems for the humid tropics', Paper presented to MAB Conference on Conversion of Tropical Forests to Pasture in Latin America, Oaxaca, Mexico, October.

——, Falesi, I., da Veiga, J.B. and Teixera, J.F. (1979) 'Productivity of cultivated pastures on low fertility soils of the Amazon basin', in P.A. Sanchez and L.E.

Tergas (eds), *Pasture Production in Acid Soils of the Tropics*, Cali, Colombia: CIAT (Centro Internacional de Agricultura Tropical).

Silveira, I.M. de and Gatti, M. (1988) 'Notas sobre a ocupação de Roraima, migração e colonisação', *Boletim Museu Paraense Emílio Goeldi, Sér. Antropol.* 4(1), 43–64.

Sioli, H. (1980) 'Foreseeable consequences of actual development schemes and alternative ideas,' in F. Barbira-Scazzochio (ed.), *Land, People and Planning in Contemporary Amazonia*, Cambridge: Centre of Latin American Studies.

—, (1985) 'The effects of deforestation in Amazonia,' *Geographical Journal*, no. 151, 197–203.

Smith, N.J.H. (1976) 'Transamazon highway. A cultural ecological analysis of settlement in the lowland tropics,' Ph.D. dissertation, University of California, Berkeley.

—— (1982) *Rainforest Corridors. The Transamazon Colonization Scheme*, Berkeley: University of California Press.

Southgate, D. (1991) *Tropical Deforestation and Agricultural Development in Latin America*. Divisional Working Paper No. 1991–20. Washington, DC, Environment Department, The World Bank.

Souza, A.F. de (1969) *Noções da geografia e historia de Roraima*, Manaus.

—— (1977) *Roraima em revista*, Boa Vista.

Souza, J.M. de (1977) *A Manaus–Boa Vista (Roteiro histórico)*, Manaus.

Sternberg, H. (1987) 'Reflexões sobre desenvolvimento e o futuro da Amazônia' in G. Kohlhepp and A. Schrader (eds), *Homem e natureza na Amazônia*, Tübingen: Geographisches Institut der Universität Tübingen.

Stradelli, Count E. (1889) 'Rio Branco. Note di viaggio', *Bolletino della Società Geografica Italiana*, 3 ser, vol. 2, anno 23, no. 26, 210–28, 251–66.

Tardin, A.T., Lee, D.C.L., Santos, R.J.R., de Assis, O.R., dos, Santos Barbosa, M.P., de Lourdes Moreira, M., Pereira, M.T., Silva, D. and Santos Filho, C.P. (1980) *Subprojeto desmatamento*, Convênio IBDF/CNPq-INPE 1979 Instituto de Pesquisas Espaciais, São José dos Campos, SP.

Taylor, K.I. (1988) 'Deforestation and Indians in Brazilian Amazonia', in E.O. Wilson (ed.), *Biodiversity*, Washington, DC: National Academy Press.

Thompson, J., Procter, J., Viana, V., Ratter, J.A. and Scott, D.A. (1992) 'The forest–savanna boundary on Maracá Island, Roraima, Brazil: an investigation of two contrasting transects', in P.A. Furley, J. Procter and J.A. Ratter (eds), *Nature and Dynamics of Forest–Savanna Boundaries*, London: Chapman and Hall.

Toledo, J.M. and Serrão, E.A.S. (1982). 'Pasture and animal production in Amazonia', in S.B. Hecht (ed.) *Amazonia: Agriculture and Land Use Research*, Cali, Colombia: CIAT (Centro Internacional de Agricultura Tropical).

Tosi, J.A. and Voertman, R.F. (1964) 'Some environmental factors in the economic development of the tropics', *Economic Geography*, vol. 40, 189–205.

Treece, D. (1990) 'Indigenous peoples in Brazilian Amazonia and the economic frontiers', in D. Goodman and A. Hall (eds), *The Future of Amazonia*, London: Macmillan.

Turner, S. (1989) 'The problems and potential use of using satellite imagery for mapping the physical growth of the city of Boa Vista, Roraima', Unpublished ms., University of Edinburgh.

Uhl, C., Clark, H. and Clark, K. (1982) 'Successional patterns associated with slash-and-burn agriculture in the upper Rio Negro region of the Amazon basin', *Biotropica*, vol. 14, 249–54.

—, Buschbacher, R. and Serrão, E.A.S. (1988) 'Abandoned pastures in eastern Amazonia, 1. Patterns of plant succession', *Journal of Ecology*, vol. 76, 663–81.

UNDP (United Nations Development Programme) (1990) *Support to the (agro)-*

ecologic and (socio)economic zoning programme for the Brazilian Legal Amazon Region, First draft, Project Document (November).

Walsh, J.J., Rowe, G.T., Iverson, R.L. and McRoy, C.P. (1981) 'Biological export of shelf carbon is a sink of the global CO_2 cycle', *Nature*, no. 291, 196–201.

Waugh, E. (1934) *Ninety-two Days*, London.

Whitmore, T. and Prance, G.T. (1987) (eds) *Biogeography and Quaternary History in Tropical America*, Oxford: Clarendon Press.

Woodwell, G.M., Houghton, R.A., Stone, T.A., Nelson, R.F. and Kovalick, W. (1987) 'Deforestation in the tropics: new measurements in the Amazon basin using Landsat and NOAA advanced very high resolution radiometer imagery', *Journal of Geophysical Research*, vol. 92, 2157–63.

World Bank (1981) *Brazil: Integrated Development of the Northwest Frontier*, Washington, DC: The World Bank.

Zambrone, F.A.D. (1988) 'Perigosa familia', *Ciência Hoje* 4(22), 44–7.

Zimmermann, J. (ed.) (1973) *Diagnóstico sócio econômico preliminar*, Boa Vista: ACAR (Associação de Crédito e Assistência Rural – Roraima).

NAME INDEX

SUBJECT INDEX